Breakfast
with the
Centenarians

DANIELA MARI

Breakfast with the Centenarians

THE ART OF AGEING WELL

Translated from the Italian
by Denise Muir

Atlantic Books
London

First published in Italy in 2017 as *A spasso con i centenari*
by Il Saggiatore, Milan.

First published in trade paperback in Great Britain in 2019
by Atlantic Books, an imprint of Atlantic Books Ltd.

10 9 8 7 6 5 4 3 2 1

A CIP catalogue record for this book is available
from the British Library.

Trade paperback ISBN: 978 1 78649 483 2
E-book ISBN: 978 1 78649 484 9

Printed in Great Britain

Atlantic Books
An imprint of Atlantic Books Ltd
Ormond House
26–27 Boswell Street
London
WC1N 3JZ

www.atlantic-books.co.uk

A still from Frank Capra's 1937 movie *Lost Horizon*, which immortalized the fictional utopia Shangri-La (© Alamy).

CONTENTS

Introduction

What does 'ageing' mean? Is it the inexorable passing of time until you catch sight of yourself in the mirror one day and don't recognize the person looking back? That moment in time when you realize the reflected you is no longer the way you see yourself in your head? But what time-spectrum are we talking about exactly? Is it the one that starts at birth but feels different as we progress through our twenties, forties, and into our sixties and beyond? Or is it the one that's shaped by individual biological and life experiences, meaning it can vary from person to person?

Ageing is a complex, dynamic and ultimately variable phenomenon. When I was younger, for example, and had to juggle family, work and study commitments, all at breakneck speed, I saw old age as a time when I would finally be able to start working my way through the stacks of books piled everywhere (the only true 'feature' in my house), take that trip I'd put off for years, enjoy lots of free time, and finally stop worrying about how I look. After many years spent studying ageing, I realize that the most common scientific definition – the accumulation of changes in cells and tissues as we grow older, bringing about the increased risk of illness and death – is over-simplified and refers purely to biological ageing.

The first time I met my publisher to discuss this book, in a pretty fish restaurant in the centre of Milan, near the university where I had taught for many years, he asked me why I had become a geriatrician. All of a sudden, the distant memory of myself as a young student flooded my consciousness, like perfume filling the air when the top is taken off. In the mid-1970s, geriatric medicine was not a compulsory part of the medical degree programme, merely a 'subsidiary' exam. With the enthusiasm of youth, I thought I'd be studying the secrets of ageing and would discover a new Shangri-La (the Tibetan valley that was home to a community of especially long-lived monks, described by novelist James Hilton in *Lost Horizon* (1933),[1] which was later adapted for the big screen by Frank Capra). But a meeting in a hospital outside Milan with the lecturer I wanted to supervise my dissertation brought me back to reality with a bump: the hospital had only one tiny ward and scores of elderly patients, many lying in corridors.

He was so busy providing patient care, not to mention exhausted due to staff shortages and a lack of funds, that he had no time for anything else. It was only many years after graduating, when I found myself studying the blood profiles of a group of ultra-long-lived individuals, that my interest in ageing was revived.

The thing that fascinated and surprised me the most was getting to know these grand old individuals and being able to explore the worlds they'd grown up in – worlds so different to the one we know today. My work in the field has taught me that ageing is a complex phenomenon, and as such requires a humanistic approach if we're to see beyond the purely biological. As my research progressed, the wider academic world also acknowledged that the demographics of the Western world had changed, as elderly adults accounted for an increasingly large chunk of the population. Geriatrics deals not only with the prevention and treatment of conditions afflicting the elderly, but also provides psychological, environmental and socioeconomic assistance. It is a discipline which is also known as gerontology (namely, the study of the biology, physiopathology and psychosocial aspects of ageing).

My experiences with older individuals and the clear need for more teaching staff in geriatrics were what made me want to be a geriatrician and choose gerontology as my field of study. It allowed me to focus both on teaching young people and on caring for the elderly. *Breakfast with the Centenarians* is the result of that decision, a choice I have never regretted.

Unsurprisingly, geriatrics is a fairly young area of study, given that we as a species have only recently been lucky enough to make it so far into old age. In the first chapter of this book, I try to convey just how fascinating the early studies of ageing were. Some of them even earned their authors a Nobel prize!

Like many of the myths surrounding the ageing process, to equate ageing with old age can be misleading – ageing begins from conception, and the second chapter explores our first years on earth.

The care aspect of my profession cannot be viewed separately from the scientific observation and research part – they are two sides of the same coin and their duality forms the backbone of this entire book, but Chapter 3 in particular. It looks at the issue of cognitive decline, which is a major concern as we grow old. Spending time with the many patients I have treated over the years, each with their own past, has shown me that cognitive decline depends on the experiences people have throughout their lifetimes. Our job as clinicians is to explore these individual histories and suggest treatments which don't always have to be pharmacological. My own research delves into such experiences, primarily looking at the importance of art, music, culture and an active lifestyle in remaining healthy in old age.

The notion that lifestyle choices can substantially affect our ageing is one of the most exciting debates in current gerontological research, the objective now being to achieve what the World Health Organization calls 'active ageing'. Chapter 4 looks specifically at this aim. I delve into new and innovative technologies that can assist us in this regard: robotics, for instance, has made the leap from the realms of fantasy to the real world as a viable research and treatment tool.

Coming full circle, in the last chapter I return to my early experiences in gerontological research, to a time in the 1990s when I embarked on the study of 'successful ageing', as it was becoming known. Centenarians, of which there are more and more in society (due to increased life expectancy and better access to healthcare in infancy, childhood and adolescence) are an extraordinary model of humanity, and a compendium of the many ageing theories proposed over the centuries.

Do centenarians hold the key to the elusive Shangri-La that we all seem to be seeking – now, more than ever?

The road to the valley is a bold and valiant one, if we are to believe Frank Capra's vision, so perhaps it is time to travel along it.

CHAPTER ONE

———•———

Ageing

Youth is happy because it has the capacity to see beauty. Anyone who keeps the ability to see beauty never grows old.

Franz Kafka

Attending Milan's La Scala one evening, I came away rather annoyed at the rest of the audience. They had hissed at the performance of *Carmen*, directed by the visionary creator of experimental theatre Emma Dante, responding to her interpretation of the opera with indignation and displaying a hankering for the more traditional productions of old. This happens whenever classics are given a makeover and unconventional productions are staged. Personally, I found Dante's work to be spellbinding, but unfortunately it sparked heated protest among others. As I came out of the theatre, I encountered a city getting ready for the festive season: the pavements were bustling, and Christmas lights and decorations twinkled over streets congested with traffic. It was a city so unashamedly artificial and far removed from the one of my youth. As I noticed this, I wondered to myself, 'Is this what old age is? This withdrawal into ourselves? This disdain for anything new and different?'

A story I'd like to share with you about a friend of mine would suggest otherwise: he is 82 years old and has an outstanding career behind him as a marketing executive at a hugely successful firm. After retiring, he attained a degree in philosophy, and he is now studying for a master's. A few months ago he invited me and a few other friends to attend a lecture by Elio Franzini, an esteemed professor of Aesthetics at the University of Milan. Our humble group – some with only a vague recollection of the subject matter from their schooldays; others who had cultivated a love of philosophy well into their adult years – accepted the invitation, curious to hear what the professor had to say. The lecture theatre was full, with young people occupying every available space, the floor included. We took our places discreetly and I felt an unexpected frisson of excitement as I watched our 'young at heart' group of genuinely mature students soak up the lesson on the Enlightenment with the enthusiasm and concentration of university students. A familiar question came to mind: 'Are we really getting old? Because if we are – and maybe we're not – when did this process that none of us is aware of begin?'

My writing this book is an attempt to answer that question. When I started it I thought, 'I'm really going to do it, I'm going to find the answer to this and to all questions about ageing.' Unfortunately, the task still eludes me. Despite a career spent studying the ageing process, it refuses to bend to my comprehension, determined to perpetually evade my grasp. I sometimes feel like I have only scratched the surface, as there is so much yet to discover.

What I have established, however, is that while medicine and biology undoubtedly help us to understand how we age, the process is so much more than the simple decline of millions of cells and genes. Every person is unique, and from the many I have worked with throughout my career, I have learned that to understand what 'getting old' *means*, we must often look beyond science for answers.

To the philosopher Arthur Schopenhauer, for example. When Schopenhauer realized his life was drawing to a close, he wrote a series of essays which were published posthumously in the form of a wonderful book called *Senilia* or *'The art of Ageing'*.[2] This short manuscript is full of anecdotes and memories, short notes and witty comments, and maxims that are hugely evocative but never tragic or commiserative. The author does not sound at all like an old man who has given up on life; quite the opposite, he is a man who continues to discuss music, mathematics, religion and Shakespeare, defending his beliefs tenaciously. He fights for his ideals and pours his knowledge and spirit into the pages of his essays, which are testament to an essential truth which is often not fully understood: as we get older, our muscles may go limp and our bones grow weak, but this does not mean that we, ourselves, become flaccid and fragile.

An interesting study of 180 novices from the American School Sisters of Notre Dame suggests a way to age with grace and dignity consistent with Schopenhauer's enlightened approach.[3] Back in the 1930s, the Notre Dame sisters had been asked to write an account of their lives, describing their most important experiences. They all mentioned key events from their childhoods, the schools they attended, their religious practices and the reason they entered the sisterhood. When researchers examined their diaries sixty years later, they found an intriguing inverse correlation between the expression of positive emotions and an increased risk of mortality in old age. Novices who had shown greater imagination and used richer language in the autobiographical accounts written in their younger years tended to live longer and were more intellectually and emotionally active than the sisters who had lacked ideas and enthusiasm in their youth.

The findings of the Nun Study are of seminal importance and can't be emphasized enough, especially in a world in which the population is ageing at such a rapid rate. The number of elderly people living in the West has grown exponentially; the same is true in eastern hemisphere nations, such as Japan. In the UK, for example, the Office for National Statistics (ONS) found that while in 1996 the number of local authorities with more than 3 per cent of the population aged 85 and older was zero, by 2016 it was over half of local authorities.[4]

This global phenomenon started last century, as a result of declining infant mortality, improved hygiene, better diet, widespread vaccination programmes and the introduction of antibiotics. But the biggest gains in life expectancy since the 1980s can be traced primarily to fewer deaths from cardiovascular disease, made possible by new medical and surgical treatments, as well as health and disease prevention programmes that have increased public awareness. This tendency will most likely continue into future decades. It is estimated, for example, that the life expectancy of a 30-year-old male would be extended by fifteen years if cardiovascular disease risk factors – such as smoking, high cholesterol, high blood pressure and obesity could be eliminated.

To get a clearer picture of population ageing in Europe, consider that life expectancy at birth has risen by an average of ten years across the European Union (EU). Of the 743.1 million people now living in the EU, 18.9 per cent are aged between 53 and 71. Italy is a perfect example of the phenomenon – according to the country's National Institute of Statistics (ISTAT), in 2015 there were 13,300,000 so-called baby boomers (the name given to the demographic cohort born in the boom years after World War Two, between 1946 and 1964, when birth rates spiked) out of a total 60,795,612 inhabitants. Similarly, in the UK in 2016, 18 per cent of the population was aged 65 and over, and 2.4 per cent aged 85 and over. These baby boomers now make up the older generations in society, and as they continue to age they pose serious challenges to health services in the developed world. It is interesting how greater longevity – living significantly longer than the average life expectancy – was the target of health policy for decades, but now that we have almost achieved it, this once-utopian endeavour may destroy the economies of many European and non-European nations.

At the start of the twentieth century, the average lifespan in England and Wales was 46 years and the main causes of death were respiratory illnesses like pneumonia, tuberculosis and complications triggered by flu. Average life expectancy has now reached approximately 81 years, and the main causes of death are chronic or degenerative diseases. This is unprecedented in the history of human ageing, and the remainder of the twenty-first century is likely to witness an even more rapid expansion of the elderly population. Clearly, this has not come about by chance. Social development, fuelled by advances in medicine, has played a key role in bringing us to where we are now.

Another surprising fact is that the number of centenarians living in Europe and other Western countries has doubled every year since 1960. Italy had only 49 centenarians in 1921, but an unbelievable 1,304 by 1981. At the beginning of 2015, ISTAT estimated that centenarians accounted for 19,095 of the country's 60,795,612 inhabitants, a number which is expected to rise to approximately 100,000 by 2050. Meanwhile, in the UK, the ONS reported that there were 14,570 centenarians in 2015, four times the number in 1985; and in France, the National Institute of Health and Medical Research (INSERM) estimated in 2016 that 21,000 citizens were aged 100 or over, twenty times as many as in 1970. Based on these figures, healthcare is set to become the largest area of government expenditure throughout much of the developed world.

———◆———

Before we dive into the unexplored waters of ageing, I'd like to start with some background. As I mentioned earlier, I've devoted a large part of my professional life to studying what it means to get older, and gerontology is very different from, let's say, orthopaedics. Injuries to the skeletal system can cause serious pain and have far-reaching effects. The body has multiple bones in a myriad of shapes and sizes, and when we break one, it takes special care and treatment to heal it. The techniques used to do this all stem from very old and well-documented principles. Gerontology, on the other hand, is a relatively new branch of internal medicine. After all, what use would ageing studies have been in the nineteenth century, when a doctor's sole concern was how to save his (mainly young) patients from the effects of war and epidemics? In a way, people didn't age – or at the very least, growing old was a luxury afforded to the lucky and rich few. Doctors had to focus on more imminent problems.

At the birth of David Copperfield, in Charles Dickens's famous novel, the sole concern at the forefront of Mr Chillip (the doctor) and Mrs Copperfield's minds is delivering the baby alive. Physicians back then had no idea that ageing starts in the womb and that everything we do in life – including before it begins – shapes our life expectancy.

The truth is that ageing is an unstoppable and inevitable process which starts at birth and ends with death. It can be scientifically defined as 'age-progressive decline in the physiological, biochemical and genetic functions needed to survive, conceive and reproduce'. The underlying causes are found in the interaction between intrinsic (genetic) and extrinsic (environmental) factors, as well as stochastic (random) events and so-called 'epigenetic' factors (a kind of mix of the above where environmental factors regulate how genes are expressed). With advances in research, we have come to realize that ageing is also the result of how all these factors interact, which makes the process unique to each individual. Of course, human life expectancy also varies from population to population: for example, a baby boy born in the UK between 2014 and 2016 can expect to live for 79.2 years and a baby girl for 82.9 years; while in other parts of the world – like Togo, Cambodia and Afghanistan – life expectancy is less than 40 years.

There's an anecdote I've always liked because of how well it captures the unpredictability of ageing. In 1965, Jeanne Calment, a French woman living in Arles, signed an agreement at the age of 90 for the sale of her apartment to lawyer André-François Raffray, a man half her age. Under the equity release scheme they used, Raffray was required to pay a monthly rent to Madame Calment until her death, at which time he would inherit the apartment. The deal seemed too good to be true, but it backfired when the elderly Madame Calment, who had the energy and mental acuity of a teenager, lived until she was 122. She outlasted Raffray, who never managed to move into the apartment he had bought. In fact, Madame Calment was the longest-living person since modern scientific records began, and her age is currently cited as the maximum human lifespan.

At every stage of our existence, even at the very end, the human body is programmed to survive, not to age and die. However, the process of natural selection drives each species to establish a hierarchy of priorities, at the very top of which are growth and reproduction. In other words, we are programmed to focus on perpetuating the species, rather than fashioning bodies capable of lasting forever. When biologist Tom Kirkwood formulated his 'disposable soma' theory in the 1970s, he drew an analogy between ageing and the industrial practice of not wasting infinite resources on increasing the lifetime of products which will be used for a limited time.[5] Kirkwood posited that, from an evolutionary perspective, we age because developing an immortal body would require too much energy. For most animals, humans included, energy is primarily earmarked for cell repair during the reproductive stage. Reproductive cells take priority, and all non-reproductive (somatic) cells are subject to deterioration and ageing.

The scientific voyage of discovery to uncover the secrets of ageing, longevity and life expectancy is, as you may already have guessed, quite an epic one. It began in 1961 when Leonard Hayflick demonstrated that there is a set number of times our cells can divide to form new cells.[6] Called the Hayflick limit, this number determines an organism's lifespan. Hayflick based his ageing theory on cultures of specific cells called fibroblasts, which are found in the skin and other tissues. He noticed that when adequate nutritional substances are present, embryonic human fibroblasts double up to fifty times before entering a senescence phase that, after a further ten cell divisions, ends with the extinction of the cell population – in other words, cell death.

Having made this discovery, Hayflick was later able to demonstrate that the number of fibroblast replications in different animal species is directly proportional to the maximum lifespan of each species (as shown in Figure 1).

Hayflick suggested that lifespan is linked to genetic factors, and that each species and each individual has their own 'inner clock' that pre-programmes the length of their existence. Writer-director Andrew Niccol seems to draw on this discovery in the 2011 film *In Time*, in which the characters are genetically modified to stop ageing at 25, when a timer on their forearm starts to count down the one year they have left to live. This time limit can be extended or shortened in different ways, and time basically becomes the new currency. It is used to pay workers, buy food, purchase a car or rent a house. Power and wealth are measured in time remaining, and the better-off have a chance of achieving immortality. In this world, nothing matters except the body clock.

Hayflick's studies may not have had the same dystopian fantasy appeal as the plot of Niccol's film, but his work holds the promise that we may one day be able to manipulate our own expiry dates. The discovery

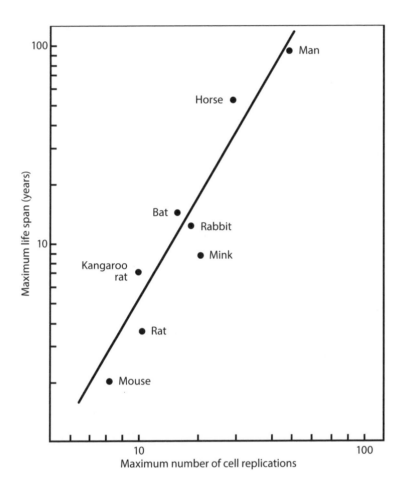

Figure I. The relationship between fibroblast replication and the maximum lifespan of an animal. (© Laura McFarlane)

of telomeres gave further credence to this theory. A telomere (from the Greek *telos* [end] and *meros* [part]) is a region at each end of a chromosome that acts like a cap, protecting each strand from deterioration. In each chromosome duplication, not all of the telomere is replicated and so it becomes progressively shorter until the cell ultimately loses its ability to divide and dies. As we grow older mutations also occur in the telomeric region. The shortening of telomeres, therefore, can be used as a way of predicting the potential onset of diseases such as cancer, and their outcome.

Not long after Hayflick's initial pioneering studies, three researchers – who would later be awarded the Nobel Prize in Physiology or Medicine for their work in this field – discovered telomerase, an enzyme which can reverse chromosome shortening and can synthesize new telomeric sequences, slowing down cell ageing. It would seem, then, that telomerase is the answer to all our ageing problems. Unfortunately, though, that's not been demonstrated in humans yet – especially as tumour cells also require telomerase to reproduce.

A group of researchers at the Dana-Farber Cancer Institute, an affiliate of Harvard Medical School, succeeded in partially reversing age-related degenerative damage in a group of mice.[7] They produced genetically engineered mice in which the gene for telomerase was equipped with a 'switch', so that it could be activated and deactivated at will. By deactivating the enzyme, they succeeded in causing very early ageing in mice. When they then reactivated the telomerase, they observed a new increase in brain and testicular mass, as well as the recovery of brain functions that had deteriorated during the rapid ageing process.

Recent data shows that telomere length and telomerase levels can also be reduced by stress; this has been seen in a sample of people who, while caring for patients with Alzheimer's – whether they were related to them or not – experienced depression. By engaging these caregivers with daily meditation exercises or getting them to listen to relaxing music, researchers were able to improve both their mood and their telomerase levels.[8]

These studies have played a key role in shaping the theories underpinning contemporary gerontological research. Please bear with me now, reader, as I attempt to explain these theories. They will help us find our way through the great mystery of ageing and will lift the veil on its many secrets.

————•————

Ingmar Bergman once said that ageing was like climbing a mountain: our muscles hurt, we want to give up, go back and have a rest, but the view from the top is so amazing that we keep going. Similarly, understanding the ageing process is a little like ageing itself; it's an uphill climb, the road is wayward and steep, yet after the initial fatigue, something precious awaits. I will try to be your guide, and I will only take you on the trickier paths when I am sure a view of outstanding beauty lies at the end.

In 1980, two American scientists, Thomas Johnson and Michael Klass, discovered that a single gene mutation could extend the life of *Caenorhabditis elegans*, a type of nematode (the so-called roundworm). Johnson optimistically named the mutated gene age-1, convinced that he was about to discover many other genes with a similar function. Although his hopes would be dashed, ten years later, the American gerontologist Cynthia Kenyon demonstrated that by removing the daf-2 gene (which regulates reproductive mechanisms), it was possible to prolong the life of this worm by 50 per cent, and this beneficial effect was even more evident in older worms. Subsequent studies have discovered another gene called daf-16, which can increase the lifespan of the roundworm by at least ten years.

Genes with the same structural features as daf-16 are present in other animal species such as the fruit fly and even in humans, whose similar FOXO3 gene is also associated with longevity. Another key factor in understanding the ageing process could be IGF-1 (insulin-like growth factor 1), a primitive hormone that was amazingly preserved in the evolution of hominids up to *Homo sapiens* and which plays an important role in the growth processes of children, promoting muscle growth even in adulthood. One of my recent studies has shown that a 100-year-old parent's child, who has an undisputed advantage with regard to ageing, has lower levels of IGF-1 in circulation than a child born at the same time but whose parents died before reaching their life expectancy.[9] This hormonal network is usually inversely correlated with antioxidant defences, so that low levels of IGF-1 correspond to an increase in the body's ability to counteract production of free radicals that accelerate ageing and cancer.

This brings me to another potential definition for the relentless phenomenon that is ageing: the progressive accumulation of unrepaired molecular and cellular damage. This makes life expectancy the length of time that a member of any species can live before the afore-mentioned damage becomes irreversible and the body enters a period of decline called 'senescence' – the end of ageing. According to the free radical theory proposed in 1956 by Denham Harman, who would later become Professor Emeritus at the University of Nebraska Medical Center, the speed of ageing is linked to the production of free radicals (unstable and highly reactive atoms).[10] All cellular reactions that produce energy also generate free radicals; their production is linked to individual metabolism, nutrition and lifestyle, as well as levels of exposure to pollution, sunlight, alcohol and smoking. Free radicals are able to damage cells by causing a chain of chemical reactions, involving oxygen. One particular site of damage is the mitochondria of cells, which are the power generators of cells where respiration takes place.

Mitochondria (specialized structures found within our cells) are fundamental for regulating multiple bodily processes. They have their own DNA, which differs from that of the cell nucleus – it is maternally transmitted and is subject to mutations more frequently than nuclear DNA. But although mitochondrial damage is evident both in physiological ageing and in many neurodegenerative diseases (such as those discussed in Chapter 3), some alterations of mitochondrial DNA are also associated positively with ageing and longevity. One of my studies on Italian centenarians demonstrated that a specific mitochondrial variant is more prevalent in centenarians than in the young population.[11] But the same variant has been associated with several diseases, so it seems that the same gene can have opposite effects, encouraging diseases in youth and longevity in old age.

One of the main regulators of free radical production is metabolism, and the only metabolic intervention that has increased life expectancy in animal experiments is calorie restriction; in other words, putting them on a diet! The results were initially convincing only in the case of rodents. In the case of primates, two major studies on macaques were carried out in the 1980s – one by the University of Wisconsin and the other by the National Institute on Aging (NIA). Only the former supported a strong correlation. However, both studies were re-examined and discussed in the *Nature Communications* journal in 2017.[12] In this reassessment, the authors highlighted differences in the two diets and concluded that it was calorie restriction (as opposed to malnutrition) that increases lifespan and postpones the onset of age-related diseases; in the NIA study, the monkeys perhaps had their calorie intake cut *too* low.

It's quite astounding how something as seemingly ordinary as reducing calorie intake can play an important role in helping us to understand the *how* and *why* of ageing. After seventy years of research, the underlying mechanisms are still not 100 per cent clear, but the data we have suggests that calorie restriction acts centrally and on a specific area of the brain (the hypothalamic preoptic nucleus), thus slowing cell division, metabolism, the production of free radicals and hormesis (an adaptive response whereby passing exposure to a toxin has a positive long-term effect on the individual's longevity). A typical example of hormesis would be alcohol consumption: drinking small amounts can protect against cardiovascular disease, whereas high-level consumption increases the risk of impaired health.

In addition to experimenting with calorie restriction, the inclusion of antioxidants in the diet became popular in industrialised countries to combat the harmful impact of free radicals, despite the lack of conclusive scientific evidence regarding their potential (positive) effect on humans. Many studies looked at what happened when you add vitamin E, an antioxidant molecule, to the diet, as a number of epidemiologic studies had suggested it could reduce the risk of cancer and cardiovascular disease. However, such observations were never proven. The effect on human ageing of antioxidant-rich diets remains unclear, and further research is required to determine if such diets affect the onset of disease and mortality. This must be done using large sample groups, followed over several years and compared with a control group not receiving the antioxidants.

Another factor to consider is inflammation, a process we will all be familiar with in one way or another. While there is much evidence demonstrating the presence of inflammation in a range of clinical conditions, its role in ageing was only recently recognized. But it is now considered a key factor in the mechanism that causes ageing, to the extent that an Italian scientist from Bologna, Claudio Franceschi, coined the term 'inflammaging'.[13]

Inflammation is a complex defensive reaction normally triggered in response to physiological or non-physiological stress, designed to recruit cells from the immune system to the part of the body in question. It has been suggested that humans have individual stress thresholds that determine how the body deals with stress, and when inflammation exceeds this level, cellular ageing occurs. This occurs via several potential pathways. Firstly, the overproduction and/or uncontrolled release of free radicals and nitrogen is very high when inflammation is taking place. Second, once in the site of inflammation, the primary function of immune system cells is to rid the body of microbes or cell debris. In so doing, surrounding tissues can become damaged, becoming harmful and resulting in the development of many conditions typical of ageing: atherosclerosis, diabetes, tumours, and neuro-degenerative diseases such as Parkinson's and dementia, to name a few.

In 2000, the theory of a defence network was developed, in which Claudio Franceschi suggested that ageing is indirectly controlled by a network of cellular and molecular defence mechanisms.[14] This communication network limits the negative effects of the physical, chemical and biological stresses that we encounter throughout our lives. The efficiency of this defence network is genetically controlled and varies between different species and individuals, which explains why life expectancy is not the same for all.

If this complex defence network fails, cells cannot maintain their equilibrium (homeostasis); they are unable to replicate and can die. It is not yet clear how each of the mechanisms involved contributes to the whole network, or how the network is hierarchically organized – especially in higher organisms. This theory combines data from cellular and molecular biology studies with the evolutionary ageing theories elaborated in the 1970s. In merging previous theories that were, by themselves, insufficient, we get a more complete view of the complex process of ageing.

A 'more complete view'. Vision and views seem to be a recurrent theme in the discussion of ageing. The views to be enjoyed from the top of a mountain, in Ingmar Bergman's words, echo the rolling landscapes of the

Shangri-La imagined by James Hilton. But does it really make sense to talk about a more complete view when the issue we are discussing – ageing – is still so shrouded in mist?

CHAPTER TWO

———◆———

When do we begin to age?

Sometimes I arise in the dead of night
and stop all the clocks, every one of them.
Yet there is no need to be afraid.
For time is also a creature of the Lord,
who created us all.

Hugo von Hofmannsthal,
The Knight of the Rose

You open your eyes. It's a beautiful summer's morning, a little after dawn. A breeze blows through the open window, a warning of the heat that will soon descend over the city. The gentle whisper brushing your still-sleepy skin feels cool, bringing a smile to your lips. You're 60 now, but on a morning like this it's easy to forget. A quick glance at the alarm clock and you're up. You jump into the shower, dry your hair, get breakfast going as you tidy the lounge and water your beloved window boxes. The smell of freshly brewed coffee beckons you back into the kitchen. You check how you look in the half-light of the hall mirror, pick up your keys and make to leave. In the doorway, you glance at your watch – an hour has passed since the alarm sounded. It's taken you twice as long to get ready as it used to. You of all people, after a lifetime of priding yourself on doing things quickly.

As time passes, signs of the transition to a slower world become more difficult to ignore – it takes you longer to climb the stairs, you need more time to read the newspaper, and where once you might have grabbed a few crackers between meetings, now you prefer to eat more leisurely, sitting on a bench and enjoying the sunshine. It is summer, after all. Then, one evening as you get home from work, you glance in the mirror and think, 'I'm getting old.'

Is this how it works? In ancient Rome you could only be appointed to the Senate once you reached 40, making it a role of undeniable seniority. Nowadays you're considered old if you can't find your way around a computer, smartphone or tablet, or if you're not a regular user of the Internet or apps. Young people – 'digital natives', as they're now called in the press and derided in academia – think that anyone not like them is old, regardless of the actual difference in age.

The more scientific response to the question of *when* we begin to get old is extremely complex. Recent studies have demonstrated, for example, that the environmental conditions we are exposed to in childhood can switch on the ageing process and the associated triggers for age-related illnesses later in life. These external conditions are the focus of epigenetics, the branch of genetic medicine that studies the chemical processes underpinning gene expression that alters the observable characteristics of individuals and any progeny, without altering DNA itself.

Epigenetics explains how the way our genes operate can change despite our genetic blueprint remaining the same. Identical twins are the perfect example. They have identical DNA, but the experience of living in different environments can result in them developing different traits (phenotypes).[15] I once studied two identical twins who grew up in different environments: despite sharing the same genetic material, one developed Alzheimer's while the other did not.

An important example of gene expression was seen in World War Two, during the Nazi occupation of the Netherlands and the terrible Dutch famine that lasted from November 1944 until spring 1945. After the Allies landed in Normandy in June 1944, they succeeded in liberating the south of the country but subsequently failed to take a series of bridges over the Rhine, including a major crossing at Arnhem. The Germans – bolstered by their first real military success since the D-Day debacle, and furious about the railway workers' strike that the Allies had called – responded by enforcing a punishing blockade on food supplies. The Dutch people suffered tremendously. The winter was bitterly cold, and communication and mobility were rendered almost impossible by the ice. A fierce battle raged for control of the region, and the retreating German army blew up bridges and river embankments to stop the Allies from advancing. The resulting floods destroyed crops, damaged farms and further impeded the delivery of badly needed supplies. In both the city and the countryside, the people lacked food and warm clothing – and very often had no form of heating either. Many were forced to eat tulip bulbs to survive. Their daily nutritional intake dropped to little more than 400 calories and sometimes even less, especially in areas where the fighting was particularly fierce.

More than 18,000 people died as a result of the 1944–45 Dutch famine, and for those who survived it, the effects were severe and reached far into the future, long after the liberation of the Netherlands in May 1945. This was shown in later years, when epidemiologists stumbled on something very unexpected when examining the detailed records that the Dutch authorities had kept during the war. They found that children born to mothers who had experienced malnutrition during pregnancy (a calorie intake of around 400) showed higher rates of obesity and cardiovascular disease in adult life, compared with the wider population. Let us examine this further.

The Dutch famine made us acutely aware of the influence of the external environment – on human health in general and on ageing in particular. The epidemiological study I referred to above, known as the Dutch Hunger Winter Families Study,[16] drew on data from the Nazi occupation of the Netherlands (an unacknowledged and tragic experiment on human health) to demonstrate how deprivation in the womb can seriously affect health in later life. The first three months of pregnancy, when the foetus is very small and growing rapidly, were found to be critical, and the effects of early deprivation were felt for the rest of their lives.

A subsequent study on subjects from the same historical cohort found that those whose mothers had been malnourished in early pregnancy showed signs of premature ageing by the time they were 56–59 years old, unlike peers born to well-nourished mothers.[17] The most likely explanation is that subjects exposed to famine very soon after conception have significantly lower rates of DNA methylation.

Let me explain that... DNA methylation is the epigenetic modification we know the most about, whereby genes are switched on or off. Situations which cause hypo- or hyper-methylation seem to activate potentially dangerous genes, or repress the good genes whose job it is to repair DNA.

The consequences of the Dutch winter can be put like this: an external factor – in this case, the hunger experienced by the mother – set in motion a set of epigenetic changes that were meant to assure the embryo would survive and still grow, despite the adverse circumstances. However, when the external environment changed – in this case, when the economy restarted after war and food was once more readily available – these 'protective' changes that were initially triggered as a survival mechanism became a hindrance under normal circumstances. The individuals concerned were more likely to experience cardiovascular problems and premature cognitive decline.

This is clearly an extreme example, yet in everyday life we can still see evidence of how people age differently. A few years ago, I attended a school reunion. Fifty years after we left our secondary school in Milan, class 3D got back together. We were all excited to see each other again, and the curiosity and fear in everyone's eyes was palpable. Helped by a visit to our former classroom, which hadn't changed much, it took just seconds for the group to rekindle the old spark. Standing within those walls, happy memories of all the years we'd spent together flooded back. It was hard to imagine that, back then, we'd all been required to wear a black smock, preferably with a white collar, and that trousers for girls were not allowed. The discipline at the school had been strict, but we all remembered how much fun we'd had with our philosophy teacher, Siro Contri. While he'd be at the blackboard with his back turned, explaining Hegel's *Phenomenology of Spirit* or whatever tract we were studying at the time, at an agreed signal we'd all move our desks up to the front or to the back, leaving half the room empty. When Mr Contri finally turned around, he'd yell, 'What's going on?! Would you all just sit still!'

We did also pick up some good habits at high school. At the end of our final year, Mr Pelosi, the Latin and Greek professor, gave us a useful piece of advice:

'Wherever you go on holiday, whether it's in the hills or on the beach, always remember to take a book with you.' I've never forgotten that.

During the reunion, we looked at old school pictures, and it was only when we compared the faces in the photographs with those sitting next to us fifty years on, at those same desks, that we realized how differently each of us had aged.

I was reminded of this observation when, not long after the reunion, I read about the Dunedin Study in New Zealand.[18] It has followed just over a thousand people (born between 1 April 1972 and 31 March 1973) from birth, assessing them at regular intervals – most recently at age 38 – on a number of physiological markers of biological age, such as blood pressure, metabolism, kidney, lung and liver function, cholesterol, immune system, teeth, and telomere length. In addition to the physiological parameters, the assessments also aim to gauge cognitive abilities like attention span, concentration, memory (patients were asked to repeat a short story), word associations, motor skills, coordination and balance.

The key strength of this study is how it incorporates a spectrum of social and genetic components which have no parallel in animal models of ageing. And the results confirm that individuals born at the same time can often age at different rates. Some of the people have shown no signs of ageing during the observation period, while others have aged a lot or had more marked signs of cognitive decline. Some even died before the most recent assessment took place. Biological age, even in young adults, appears to be highly variable.

This shows us that we also have to look at how ageing advances in young people; our research should aim to target the broader ageing process rather than focusing on the individual conditions it brings with it. In this regard, the Dunedin Study has opened up fascinating avenues of scientific exploration and potential new areas of study. We could look, for example, at how traumatic experiences like child abuse might accelerate ageing in later life, or explore how social inequality affects health. And the political, social and scientific implications are enormous. If we were to discover, say, that children born into poverty age much faster than those born into wealthier families, we would have to ask ourselves how to reverse this predisposition to premature ageing triggered in childhood – but also how to end the social inequity causing such early ageing. Moreover, using the latest methods of analysing DNA, scientists could compare gene expression levels in an individual with impaired health to those of a healthy individual, and as a result pinpoint the specific genes involved in the particular pathology. Such technology would help to diagnose early onset ageing, and aid in the design of prevention strategies to put in place before age-related pathologies emerge.

It is longitudinal analyses on large sample populations that bring such possibilities within our reach – indeed, prospective cohort studies that follow large groups of people over several decades are the only ones that enable us to identify risk factors for a range of pathologies and understand what to do to prevent them. The Framingham Heart Study, carried out in Framingham, Massachusetts, also conducts this kind of seminal research.[19] Looking primarily at the risk of heart disease, it signed up its original cohort of 5,209 subjects – aged between 30 and 62 – in 1948, engaging them in lab assessments and medical check-ups every two years. By using sophisticated statistical techniques to compare parameters measured at the beginning of the study with pathologies in later life and causes of death, a number of key trends have been identified. These include the effects of smoking, cholesterol levels, high blood pressure and obesity – conditions which affect all social classes.

This ongoing project offers anyone working on new epidemiological research access to its enormous database. It is also expected to continue to publish new and important findings.

———◆———

The example of the Dutch famine earlier in this chapter is clearly important but trauma and social inequality are not the only factors that are instrumental in the ageing process. Maintaining a positive self-image and adapting behaviour in response to physical changes can be critical in slowing down ageing. Grazia Deledda, an Italian writer who received the Nobel Prize for Literature, provided an emblematic account in her 1910 novel *Sino al confine* (*Up to the Limit*), of how being happy with oneself is an effective means of defence against the ravages of time:

> '*How are you, Uncle Bustia? Do you remember me?*'
> *The man stood up and put out a hand, rubbing it clean on his cloth trouser leg first. He hadn't aged – he had a calm, solemn air, his head gleamed, his beard was well-kept, he looked not like an old man struck by ill-fortune and dishonour, but a patriarch, proud of both himself and his descendants.*

Recently, a group of researchers at the University of Massachusetts carried out an in-depth study of 242 adults aged between 40 and 95, looking at how baby boomers were ageing. They were particularly interested in self-esteem.[20]

The demographic boom in the years after World War Two – which saw 75 million births in the United States alone – changed society in a way that was unprecedented in the history of mankind. By the end of the 1940s, families were moving out to the suburbs, sparking the construction and expansion of whole new neighbourhoods. By the time the children of these families became adults, their tastes had been profoundly changed by widespread economic prosperity, and these quickly dominated trends in the rest of the world. The baby boomers became the punk generation, the political activists, the protestors of 1968. They became the people who, in the 1980s and '90s, were responsible for putting consumer mania into capitalism. They wanted everything big – big houses, big cars – and were generally more concerned with how they looked. The University of Massachusetts study found that people in this cohort were obsessed with staying young-looking for as long as possible. Unsurprisingly, the early nineties also saw the rapid expansion of a cosmetics industry that targeted women *and* men, probably inspired by the desire for eternal youth of all those born between 1946 and 1964, some of whom were by then nearing retirement.

The study also brought to light that not all baby boomers are the same. Some people have a strong enough self-image not to be phased by remarks about greying hair or extra weight they're carrying. Such individuals don't think of themselves as 'old', and their sense of identity does not change as they age. Rather than worrying about wrinkles and grey hair, they're more concerned about maintaining their independence, their hearing, their balance and being able to get around. People with high self-esteem in later years are also able to keep their anxieties and concerns at bay, even when dealing with serious health conditions on top of normal age-related changes.

It is now universally recognized that, whatever your age, 'feeling young' helps to keep the brain young. A group of researchers from Purdue University in West Lafayette, Indiana, conducted a ten-year study of 500 people aged between 55 and 74, most of whom had stated at the beginning of the study that they felt younger than their chronological age (the average difference between actual and perceived age was twelve years). The study found that those who felt younger were also more confident about their cognitive abilities.[21]

How we perceive ourselves at a given point in time – namely, how we experience ageing – can have a positive effect on how we age in the future. Referred to as 'age identity', it has been found to be more important than actual chronological age in determining quality of life. Researchers believe that mental age can even influence cognitive ageing and slow down the process of cognitive decline.

A colleague once sent me an 83-year-old woman, and asked me not to tell her that she was about to undergo a geriatric examination. If she had known the truth, the woman, who had been a secondary school mathematics teacher for many years, would have refused to come. On the day of the appointment, a strong-willed, confident woman walked into my office. She was smartly dressed and had recently returned from a trip to the Middle East which had left her with a nasty gastrointestinal complaint. That was why she'd gone to see her doctor. Her daughter was with her, but she was shushed by her mother whenever she tried to intervene to better explain the medicines her mother took. In actual fact, tests showed the mother's memory to be excellent; on a numerical reasoning test (continuously subtracting 7 from 100, until you reach 65) she was quicker than some volunteer student 'patients' I use at lectures to show how the test is performed! On the whole, the woman – who had been happy to undergo a general check-up but not a geriatric one – did not see herself as old, and in cognitive terms she most definitely wasn't.

The Gerontologist, the official journal of the Gerontological Society of America, published a special issue in February 2017 with an unusual title – 'Aging: It's Personal'. The articles in this edition were written by nineteen academics who had been asked to think about how their knowledge influenced their own ageing experience and that of their loved ones. The editorial stated:

[…] we challenged authors to connect their personal experiences with gerontological literature, addressing questions such as 'Where were personal experience and scholarly research in conflict?' 'How has your aging experience been affected, for better or worse, by your knowledge of the literature?' 'How can your personal aging experiences inform the field?' 'Are personal experiences helpful or harmful to gerontologists?' 'How can knowledge from personal experience inform new research and theory about aging?' 'How can personal experience guide interventions aimed at improving the lives of older people?' We asked that essays be part memoir, part practical guide, and part prescription for change; that they be provocative, but

scholarly; and that they enhance or refocus the way we think about aging.[22]

Our own experiences can help us to fill the divide between theory and practice that often opens up in gerontological research. For example, when faced with a family member suffering an unexpected state of mental confusion, the knowledge and scientific guidelines we would normally call on in our professional lives are often of little help, while our personal experiences can bring new insights to the research domain.

A few years ago, my mother suffered sudden-onset hallucinations and delusions of persecution, and no amount of the regular medication used in these cases would calm her down. Worse still, she was the most aggressive with me, the one person who should have been able to treat her. I was very upset and at a complete loss, until a week later, going through her various boxes of pills, I came across an antibiotic she had been taking for a simple bout of cystitis. I hadn't prescribed it. The patient information leaflet described the various side effects, one of which was possible hallucinations in the elderly. As soon as she stopped taking it, my mother was once again the cognitively healthy woman she had always been. We doctors should always remember to check the medications patients are taking before prescribing anything which could be either ineffective or potentially harmful.

One of the reasons I decided to write this book was to enable me to step outside the closed system of the research world and draw upon my many years of knowledge – acquired not just from academic studies and analysis, but also from meetings and conversations, colleagues and patients, battles both won and lost – and my own personal experience.

The book came about after I received a letter from the university – not that long ago – advising me in the cold, hard language of officialdom that in eight months' time I would be 'retired from service'. I must admit it took me a little by surprise. I was still working pretty much flat out, splitting my time between teaching several degree courses, heading up the postgraduate programme, treating patients and doing research. The months flew past, and in that time I managed to secure funding from the EU's Horizon 2020 programme, and the publisher of this book asked me if I wanted to write about ageing in a way that combined a scientific perspective with insights from my clinical and personal experience. Life had seemed to stop when that letter arrived, but thankfully it took off again just as quickly with a whole new set of exciting challenges. It was as if I could hear the buzzing of neurons firing in unison in my brain as they set to work once more.

It doesn't surprise me, then, that people turn to geriatricians for help with melancholy partners who, in retirement, have lost interest in everything and do nothing but sit and watch television or play on a device all day. Being retired means being at home more, and this can often upset the balance of a relationship. Instead of finding each other again as a couple, nerves can become frayed and tempers lost. On numerous occasions in the academic world, I've seen former colleagues wandering the corridors and lecture halls of the department, years after retiring, and I've often wondered if there were partners at home somewhere heaving a huge sigh of relief that they'd managed to get them out of the house for another day.

In Japan they have an official condition called 'retired husband syndrome', the symptoms of which are stress, insomnia and depression... in the man's wife. This condition is not unique to Japan though, as the rest of the world has experienced it too.

I was working on this book when I happened to watch a German film that parodied the 'Hannibal is at the gates!' chant the citizens of ancient Rome repeated as they awaited the arrival of their attacker – but this time the doom on the horizon was retirement. Although the film had a happy ending, it reflected the reality that retirement is a difficult period of life – the feeling of being suddenly old can often be overwhelming, especially if one's working life comes to an end more abruptly and much earlier than expected, something that's a regular occurrence in times of recession.

When this happens, time seems to stretch out endlessly before us, and the fact that we don't know how to enjoy being idle doesn't help. As Bertrand Russell reminds us in *The Conquest of Happiness*, 'To be able to fill leisure intelligently is the last product of civilization, and at present very few people have reached this level.'[23] And recent scientific research supports Russell's position, demonstrating that stopping work and embracing our retirement years is often difficult to do, but can be good for both mind and body if the time is filled with fulfilling activities. A study published in the *British Medical Journal* followed more than 11,000 men and 2,000 women working for France's national gas and electricity supplier, assessing them annually over a period of fifteen years (the seven years before retirement and seven years after).[24] It emerged that one of the benefits of retirement is feeling less tired, both mentally and physically, and experiencing fewer feelings of anxiety and sadness. In many cases, years of work-related stress and a hectic lifestyle can have an unhealthy effect, and retirement is a welcome cure. Free from work, people can spend their newly acquired time on pursuits that were previously neglected – things like reading, going to the theatre, cinema, exhibitions, and doing more exercise.

A study known as the Nurses' Health Study, started by Harvard professor Dr Frank Speizer in 1976 and now into its third generation, followed over 100,000 American nurses across a number of years. They were given a questionnaire and asked to subjectively assess their quality of life at the time of retirement. They were interviewed a further three times – the first time one month after retirement, and the last time five years after – to assess their cognitive function. The women who reported an improved quality of life after retirement (61 per cent) were found to have a small but significant difference in cognition compared with those (31 per cent) who felt their quality of life had worsened. (The remaining 8 per cent perceived no difference before or after).[25]

———— ◆ ————

In an interview in May 2017, soprano Renée Fleming announced that she would be singing her final performance in the role of Marschallin in Richard Strauss's *Der Rosenkavalier*, directed by Robert Carsen at New York's Metropolitan Opera. The interview cited lines from the libretto: 'Time is a strange thing. While one is living one's life away, it is absolutely nothing. Then, suddenly, one is aware of nothing else.'[26] Such a sentiment echoes Oscar Wilde when he wrote, 'The tragedy of old age is not that one is old, but that one is young.'

Although science may be able to tell us in biological terms when exactly it is we begin to age, quotes like these illustrate beautifully that there is something subtler which inevitably triggers the ageing process – something that no study, no matter how big, can ever stop. For some, ageing does not start when we see our first wrinkles; it's when we give up on our dreams, when we swap true love for something safer, more suitable, and when we lose interest in others and in everything new. Unfortunately, we don't all have a portrait in the attic like Dorian Gray, to show us when and how much we are ageing inside – 'The picture, changed or unchanged, would be to him the visible emblem of conscience.'[27] Without such an emblem of our own, we need to make sure we don't let the stress of everyday life get on top of us, nor prevent us from finding the time to stop and think about how our lives are progressing. As Albert

Schweitzer said, 'The tragedy of life is what dies inside a man while he lives.'

We can always hope to make amends for times when we might have been selfish or indifferent to the feelings of those close to us. Time denied to our children, for example, can be redeemed in the event of a happy occasion like the birth of a grandchild, or if we find ourselves in other, unexpected circumstances, as dramatically portrayed in the film *Tenderness* by Italian director Gianni Amelio. When an elderly lawyer, estranged from his children on account of his many unethical and shady dealings, witnesses a tragedy that befalls his neighbours, he decides to try to rebuild relationships he has long neglected with his family. But our aim should be to not become like the old man described by Italo Svevo, angry and egotistical:

> *To be old for the whole day, without a second's reprieve! To age constantly, in every second! I struggle to inure myself to how I am today, and yet tomorrow I subject myself to the same labour, to ensconce myself in the chair which has become more uncomfortable, even than yesterday. Who would deny me the right to talk, to shout, to protest? To protest is, after all, the quickest route to resignation.*

The nineteenth-century Italian philosopher Giacomo Leopardi reminds us that 'the habit of resignation... always gives rise to heedlessness, negligence, indolence, inactivity, and finally to laziness, sluggishness, and insensitivity, and almost to immobility'.[28] Immobility is one thing we should most definitely avoid if we want to embrace ageing with the right spirit, and I don't mean just physical immobility. I mean immobility of the mind – withdrawing from all that is new and exciting around us every day. As Carla Mazzola, an elderly ward sister I once worked with, regularly pronounced when something high-tech would be introduced to the ward: 'What a pity we have to die!'

CHAPTER THREE

——•——

Cognitive decline

Just as iron rusts from disuse, and stagnant water putrefies, or when cold turns to ice, so our intellect wastes unless it is kept in use.

Leonardo da Vinci, *Codex Atlanticus*

Sister Carla was right: what a pity we have to die. There are new joys to behold every minute. Yet there are times when old age is such a burden, we forget how good it feels to discover something new – whether about ourselves or about the world. There are times when old age can even make us forget our own identity.

I remember a man who came in to see me once. He sat in silence while his daughter blurted out a breathless explanation for their visit. 'Dad has always been very active. He worked eight, nine, sometimes ten hours a day. We'd all have dinner together when he got home, and we'd listen to all his stories about the day – meetings, the problems he'd dealt with, the business lunches. He was so dynamic and cheerful, and he loved to go hillwalking and play sport. He liked to read, too, and I remember him sitting there in his chair, glasses perched on his nose, a book always open on his knee. He could recite by heart the poems he had learned at school, whole verses from his favourite, *The Divine Comedy*. But seeing him now...'

Then the woman stopped and turned to her father. I could see everything I needed to know in that look. I saw that the wonderful, learned, keen-minded paterfamilias that she had just described had ceased to exist, lost to periods of temporary disorientation, loss of memory, and apathy. As the daughter spoke and grew increasingly upset, the man sat in silence, staring into space. I asked him a few of the easier questions from the first screening test used for patients showing symptoms of cognitive impairment. It's called the Mini-Mental State Examination (MMSE), and it examines orientation to time and place, the ability to register and recall a series of prompts (*apple*, *penny*, *table* ...), attention span and calculation, language, and constructional praxis (building, assembling and drawing objects). In practice, patients scoring between 18 and 24 out of 30 are considered to have an initial form of impairment.

I asked the man in front of me, who was still staring blankly at a spot behind my head, to write a simple sentence containing a subject and a verb – a classic primary school task. I could hardly believe it when I looked at his answer: *Old age is so frustrating.*

My patient showed clear signs of cognitive decline, but he was still aware that something wasn't right and had an opinion about it. The trademark perspicacity his daughter now mourned wasn't completely gone.

I have dealt with numerous cases of cognitive decline in my career as a geriatrician. They are often the ones that require extra care, not just because the symptoms are more common in old age, but because the treatments currently available may not be successful and the impact of the condition on the wider family must also be considered. Cognitive decline affects 35 million people worldwide. In the UK, dementia (caused by Alzheimer's and other illnesses) was the primary cause of death in 2015 (11.6 per cent of 529,655 deaths), putting it ahead of cardiovascular disease for the first time.

The most famous form of cognitive decline, Alzheimer's disease, is named after a German neurologist, Alois Alzheimer. While working at a psychiatric hospital in Frankfurt in 1901, he observed a 50-year-old female patient named Auguste Deter, who had been referred to him for memory and language problems, disorientation and hallucinations. Her husband also said that she had been having behavioural issues. She was aggressive at times and also careless with her appearance. Alzheimer was surprised to see these symptoms in a relatively young woman. He followed Deter throughout her progressive decline, which ended in her death five years later. Alzheimer conducted the post-mortem and studied the results in the Munich laboratory of another influential German psychiatrist, Emil Kraepelin. Examining the deceased's brain tissue under the microscope and using special chemical dyes, Alzheimer identified insoluble and non-degradable protein deposits that had compromised the function of Auguste Deter's brain. But it was only in 1910, four years later when Alzheimer presented the results of his research to the scientific community, that Alzheimer's and dementia were first named.[29] A century later, his findings still inform current research. Indeed, in 2012, *The Lancet Neurology* published news that a team of researchers had discovered – in samples Alois Alzheimer had prepared from Auguste Deter's brain – a mutation in the presenilin-1 gene linked to some familial forms of early-onset Alzheimer's.[30]

Hereditary forms of the disease, caused by the transmission of an abnormal dominant gene, are extremely rare and account for 3 per cent of cases. But sporadic Alzheimer's is the most common cause of dementia in elderly people living in the West, and women are the most affected. While the percentage of men suffering from this form of Alzheimer's has nevertheless increased slightly, women are still three times more likely to get it, in every age group: there are around 300 cases in every 100,000 women at age 60–69; 3,200 in every 100,000 at age 70–79; and 10,800 in every 100,000 over the age of 80. Age is believed to be the highest risk factor, followed by gender, and then cardiovascular damage. However, we don't yet know the specific cause of sporadic, non-hereditary Alzheimer's. It is believed that some environmental factors (yet to be identified) interact with a genetic predisposition to bring on the illness.

Alzheimer's is a progressive disease that can last twenty years and gradually destroys brain cells. It causes an irreversible deterioration of all higher cognitive functions, such as memory, reasoning and language, until the patient is no longer able to perform everyday tasks or function independently. The progression of the disease is not the same for everyone. Originally, it was thought that it was down to the level of education achieved by an individual, which links to the hypothesis we will discuss later that culture, intellectual curiosity and artistic expression are important factors in coping with ageing. However, as life expectancy and the average age of the population increases, many highly educated artists, university lecturers and entrepreneurs with a lifetime of learning behind them are also diagnosed with dementia. The upshot is that the disease most likely emerges in advanced old age, when the cognitive reserve built up through education and learning can no longer offset degenerative brain changes.

I once treated a famous architect, internationally renowned for his bold designs. It was distressing to watch him fumble with a pencil, bewildered, drawing nonsense marks on a sheet of paper during a cognitive test. When asked to copy two intersecting pentagons, the result bore little resemblance to the original (see Figure 2).

Figure 2. A dementia patient was asked to reproduce a simple diagram. (© Daniela Mari)

There is substantial variation in the onset of dementia. The first sign that an individual might be developing dementia is a loss of interest – and progressive memory loss – in all things related to their professional life, as seen in a manager I once treated. Incredibly, as this man's professional identity disappeared, the musical composition skills he'd abandoned in childhood returned – and quite admirably, too. In the early stages of the disease, patients may report mild memory lapses, in particular regarding recent experiences, or difficulty orientating themselves to time and space. They may no longer be able to find a familiar place, even the street they live on, or they may forget important dates. After years of working in the field of geriatrics, I still find it heartbreaking to watch the terror on patients' faces when they realize they are losing their memory and as they wait for the diagnosis they fear is about to come. I then have to watch them progressively lose all sense of what is happening to them (disease insight).

A brilliant Julianne Moore interprets this 'losing oneself' in the 2014 film *Still Alice*, directed by Richard Glatzer and Wash Westmoreland. Adapted from the novel by Lisa Genova, it tells the story of a 50-year-old linguistics professor at Columbia University who is struggling to come to terms with devastating bouts of forgetfulness, which are diagnosed as early-onset familial Alzheimer's. And the same intensity, the same struggle, the same devastating illness is at the centre of *Away from Her*, a 2006 film by Sarah Polley. Starring an outstanding Julie Christie, the film is based on the short story 'The Bear Came Over the Mountain' by Alice Munro, from the 2001 collection *Hateship, Friendship, Courtship, Loveship, Marriage*. The blanket of snow covering the fields around the protagonist Fiona's house, which creates an infinite and endless sheet of silence, is a metaphor for the terrible poignancy of Fiona's disappearance into herself, and for the gradual dulling of her senses that we – the audience – and her husband, Grant, can only stand by and watch as powerless bystanders.

One unique case of Alzheimer's worth a mention is that of the American artist William Utermohlen.[31] His final series of self-portraits, which he started painting when he was first diagnosed in 1995, at age 61, are a unique documentation of the painful descent into dementia that preceded his death in 2007. The artist fought to stay in contact with the world around him, but his paintings became flatter, more abstract, with a progressive loss of detail and depth. After he was taken into an elderly care home in 2000, Utermohlen's memory and drawing skills deteriorated to the point that his final, heartbreaking, self-portrait is simply a rough sketch of a skull with a few facial features in shadow.

——•——

So far, we've talked a lot about Alzheimer's disease, and while it's the most common form of dementia, it's not the only one we see in clinical practice. Another type is known as 'ischemic vascular dementia', which is one of the first diagnoses to be suspected if the patient has a clinical history of cerebrovascular episodes (for example, strokes or transient ischemic attacks – mini strokes also caused by lack of blood flow to the brain). This form of dementia follows a more 'stepped' progression than Alzheimer's disease, characterized by sudden changes in cognitive ability followed by periods of relative stability.

Frontotemporal dementia is a third type of dementia, which develops as a result of a degeneration of the neurons located in the frontal and temporal lobes of the brain, the causes of which are unknown. At its onset, it is usually characterized by behavioural disorders (disinhibition), alterations in mood, disinterest in loved ones, hypochondria and monotone and persistent talking, often using obscene or embarrassing language. I saw one such case when I met a couple who were going through a deep crisis. The husband – a forlorn-looking man, perfectly dressed, with his hair parted dead straight as if with a ruler – came to my office sadly reporting a change in his wife's behaviour. She was known to be very reserved, but had suddenly started to behave immodestly and inappropriately. Even during the medical examination the woman continued to use vulgar language, winking brazenly at my young male colleague who was in attendance.

Finally, there is Lewy body dementia (the second most common type of progressive dementia after Alzheimer's), which owes its name to the neurologist Friedrich H. Lewy. In 1912, he discovered protein deposits in the cortex and brainstem of deceased patients with Parkinson's disease. These clumped particles, referred to as 'Lewy bodies', affect neurotransmitters (substances that help transmit signals from one nerve cell to another), thereby damaging neurons. The causes of this kind of dementia are unknown, and it affects men more than women. It frequently starts with hallucinatory episodes that are typically visual and psychotic; it also features poor thinking, lack of judgement, and individuals' sleep is often disturbed by violent movements. Such symptoms are very similar to the dementia associated with Parkinson's disease, making it difficult to distinguish between the two in diagnosis.

As patients often live quite happily with their visions and don't experience any memory loss to begin with, it can be difficult to convince them there is something wrong. They will insist on doing everything as usual – driving, for example – much to the dismay of their families, who are forced to come up with a series of ploys to steer them away from anything which may put themselves or others in harm's way.

———— • ————

Although various forms of dementia exist, their sufferers display the same emotions – fear, apprehension, confusion, anxiety, danger – and I see these every day in my job. It is crushing to see a loved one change and lose themselves, no longer able to recognize their wife, husband, child or brother.

As the Roman poet Juvenal once declaimed: 'But worse than any loss of limb is the failing mind which forgets the names of slaves, and cannot recognise the face of the old friend who dined with him last night, nor those of the children whom he has begotten and brought up.'[32]

The prospect that this could happen to any of us is terrifying. And it's the reason many pre-geriatric patients, spurred by the fear of Alzheimer's and the realization that they often can't remember where they've put their keys or their glasses, seek a consultation with their doctor in order to check for signs of cognitive decline.

The neurobiologist and Nobel laureate Rita Levi-Montalcini's words come to mind here: 'If you can't remember where you left your keys, don't immediately assume it's Alzheimer's; start worrying when you can't remember what keys are for.'[33] Researchers in the late 1990s, aware of the importance of early diagnosis, introduced the idea of mild cognitive impairment (MCI).[34] MCI is when someone has problems in one or more cognitive areas (like memory, attention span or language), as confirmed by specific neuropsychological tests, but not to the extent that the impairment might disrupt their normal, everyday life. It is estimated that, every year, between 10 and 15 per cent of people diagnosed with MCI progress to Alzheimer's. This is why, in the scientific world, a lot of our research focuses on identifying warning symptoms of the disease. If specific tests could then pick up on these markers and enable early detection, we might be able to introduce treatment in line with the diagnosis.

Symptoms very often remain unchanged over short periods, and an impairment can actually get better after the cessation of stress, for example, or if alcohol or substance abuse is tackled. But while stress and substance abuse can be addressed, the overall trajectory of the disease is unrelenting. For this reason, scientists must plough on with their research: we need to find the causes, learn more about the progression of the disease and trial new treatments. There is a lot we still don't know about Alzheimer's and dementia more broadly.

According to the most solid theories on which available therapies are based, Alzheimer's is characterized by the presence in the brain of numerous amyloid plaques and neurofibrillary tangles. These are clumps of protein produced by the brain that accumulate rather than being eliminated naturally (as they would in a healthy individual).

The precursor of the amyloid protein – the amyloid precursor protein (APP) – is a large membrane protein that performs various functions when intact, and other functions when it is reduced to fragments. Normally, it plays an important role in the growth and repair of neurons, but it turns into something harmful in its shortened form. The fragments cause neurons to die, and it is believed that this harmful action begins many years before memory disorders appear. The area of the brain most involved seems to be the hippocampus, which is responsible for the processes of learning and memory, but neuronal death gradually extends to other areas of the brain too.

At present, scientific research aims to offer the most accurate diagnosis possible using the means currently available, such as MRI scans to determine predisposing genetic markers that in turn differentiate between the different types of dementia. But, as I was saying earlier, Alzheimer's is far from a fully solved mystery. No matter how much we study it, something still escapes us. It seems tragically ironic that a disease that erodes memory and destroys personality can be so elusive, so difficult to pin down. Yet the danger is real, for all people, in every country in the world, and age is the only risk factor of which we can be absolutely certain. Age is also the reason women are more susceptible – we live longer.

It is believed that the shortcomings of the treatments used so far – which consist of drugs that can improve the symptoms of the disease but not interrupt or slow down its relentless progression – is in part attributable to a late diagnosis, often made many decades after the onset of the first symptoms. But the idea of screening all middle-aged subjects for the first deposits of beta-amyloid (the main component of amyloid plaques) is, at the moment, a project that would prove to be too expensive on a large scale. It would require neuroradiology imaging with a specific tracer – a very expensive radioactive drug that degenerates rapidly. This means that, if the examination has to be postponed, the drug is no longer usable.

Another method of screening can be invasive: it requires cerebrospinal fluid to be collected through the spinal cord in a procedure known as a lumbar puncture. It is therefore crucial – and this is one of the most interesting aspects of current research in the field of neurodegenerative diseases – that we commit ourselves to finding biomarkers for dementia that can be identified by means of a simple peripheral blood sample. Until recently, none of the tested biomarkers had proved to be significantly sensitive or specific. However, promising research from several teams in recent years suggests that simple blood tests for the presence of beta-amyloid deposits might only be a few years away.

In January 2014, then President of the United States Barack Obama signed a funding bill containing a record-breaking $122 million increase for Alzheimer's research. Previously, in 2012, he had stated his administration's goal to prevent and effectively treat the illness by 2025. The limitations of existing treatments, when considered alongside our growing knowledge of the development mechanisms of Alzheimer's disease, have steered research towards new therapeutic approaches.

As a result, trialling has recently begun on new drugs designed to affect the abnormal deposits of proteins in the brain that alter communications between synapses and cause the death of neurons. The new treatment aims to trigger an immune response in the brain to such deposits, hence modifying the natural progress of the disease, blocking or slowing it down – or even preventing or delaying its onset.

The other current focus of therapy is the tau protein, the accumulation of which leads to the formation of neurofibrillary tangles and thus to the destruction of the neurons of patients with Alzheimer's disease. So far, the results obtained with anti-tau drugs have not been convincing, but experimental studies are under way of drugs that block the beta-amyloid precursor. If this therapy were to prove successful, the amyloid substance would not degenerate into neurotoxic fragments.

There are other promising directions of research. In May 2017, an authoritative editorial was published commenting on a study that showed that the plasma of the human umbilical cord is able to revitalize the hippocampus and improve cognitive function in aged specimens of laboratory mice.[35] In particular, a protein that is plentiful in human umbilical plasma, and in the plasma and hippocampus of young mice, improves synaptic plasticity and the cognitive performance of aged mice once it has been systemically transfused. These results are the first to confirm that proteins present in human blood can improve age-related problems in mature mice. Scientists believe that the TIMP2 protein, when injected, has positive effects on the hippocampus, reactivating the mice's instincts to build a nest, a function that declines progressively with age. Identifying a specific protein that can provide these benefits is a good starting point for the future development of pharmacological treatments for Alzheimer's.

———•———

Although Alzheimer's and other forms of dementia are frightening prospects, and scientific research on the subject is often inconclusive, it's important not to forget that there is life beyond the disease. Even in its most advanced stages, we shouldn't give up, lay down our arms and surrender to the disease. On the contrary, this is the very time we need to fight and make sure that life prevails.

Two psychoanalysts, Maddalena Muzio Treccani and Mario Rivardo, led a fascinating experiment at a nursing home in Milan.[36] I headed the geriatric unit there for three years, and the two researchers set up an art lab specifically for patients afflicted with Alzheimer's and dementia, some of whom also had motor impairments. The results of the lab took family members by surprise, and many broke down in tears at the artwork their loved ones had produced. Until that moment, they'd believed them 'lost', locked in a world that their family no longer had access to. Looking at the use of colour, the words added to the drawings and the communication established with the patients, the psychoanalysts saw signs of the psyches that lay within them, even at such an advanced stage of the illness. The world in which the patients had seemed lost was, all of a sudden, much closer than they'd imagined.

In a patient report, when describing the psychological evaluation, Muzio Treccani concluded: 'the coloured threads in the drawing are memories [...] weaving together the stories that Rina narrates and reinvents within the context of her dressmaking shop, at times rich and full of grace, at others full of drama.'

When I read about this, it reminded me of one of the tales in *The Man Who Mistook His Wife for a Hat*, a book of case studies by neurologist Oliver Sacks:

Complementary to any purely medicinal, or medical approach, there must also be an 'existential' approach: in particular, a sensitive understanding of action, art and play as being in essence healthy and free, and thus antagonistic to crude drives and impulsions, to 'the blind force of the subcortex' from which these patients suffer. The motionless Parkinsonian can sing and dance, and when he does so is completely free from his Parkinsonism [...] here the 'I' vanquishes and reigns over the 'It'.

Individuals adrift in the murky waters of Alzheimer's can surface, even if only briefly, through art, painting and music. In a 1916 poem of the same name, the Italian modernist Giuseppe Ungaretti wrote of arriving at a 'buried harbour' and then returning to the surface with new songs, which echoes how artistic expression can bring back to light the true essence of a patient afflicted with Alzheimer's; the essence that the disease seemed to have permanently destroyed.

A Canadian occupational therapist, Moyra Jones, adopted a similar approach when caring for dementia patients.[37] She wanted to improve her patients' well-being by removing any potential stress from their lives. To do this, her team focused their attention on the physical space, striving to provide an environment that matched their life histories and personal identities, and contained objects that would help them to understand the function of their surroundings. This included familiar hats, bags and clothing, newspapers and books, and glimpses of the outside world through the large windows overlooking the facility's garden or surrounding areas. We know that being close to the natural environment is important in the treatment of neurological disorders.

In the rehabilitation facility where Jones worked, the daily routine was built around the needs of patients, not the staff. This was done in an attempt to create, as far as possible, an environment that reflected the life of the individual and their remaining functional capacities. Occupational therapy and any specific abilities the staff were able to share in (music therapy, painting workshops, rehabilitation exercises) were used to cultivate a patient's strongest passions. In this context, a working relationship between staff and family members or primary caregivers is vital, given that the care provided is flexible and must often be adapted as the patient progresses through different stages of the disease.

The results achieved by Moyra Jones and other like-minded scientists are extremely encouraging, making it possible to limit patients' reliance on drugs and methods of restraint – thus reducing the stress on patients and their caregivers when dealing with behavioural disorders.

———— ◆ ————

Caring, supporting and treatment are only one side of the story, though. As is the case with other complex disorders influenced by both environmental and genetic components, Alzheimer's research is currently concerned with finding effective prevention strategies. We have already touched on the importance of the cognitive reserve, namely the skills and knowledge we acquire over the course of a lifetime, not just in early childhood. With a greater cognitive reserve, we are better able to stave off degenerative brain changes.

Studying, learning, understanding, travelling, visiting a museum or an exhibition, reading – these can all help to build up a crucial resistance to diseases which cause damage to the brain. Making changes to the cultural component of our lifestyles can therefore have a positive effect on cognitive health; and with this in mind, scientists have long been investigating which other aspects of our lifestyles might be adapted to similar effect.

Diet is one such area. The consumption of meat and mature cheeses, rich in AGEs (advanced glycation end products), has come under attack because, according to a study carried out on large patient cohorts, these foods are a risk factor in the development of Alzheimer's.[38] AGEs are the result of sugars combining with protein or fat. They are a normal feature of ageing, but are also introduced into the body through diet (meat cooked at high temperatures, fried food, processed grains, cheeses). AGEs are reported to increase the risk of Alzheimer's through multiple mechanisms – increased cellular inflammation, neurodegeneration and oxidative stress. The study recommends Mediterranean and Japanese diets, which reduce the build-up of AGEs due to lower meat content and use of a wide variety of vegetables and grains. Precooked foods or ready meals, in which AGEs regularly occur within additives to enhance colour and flavour, are to be avoided.

In a recent study, known as the Northern Manhattan Study, 876 participants of an average age of 71, with ten years of formal education, were given a set of neuropsychological texts to examine their memories, and then were tested again five years later.[39] The participants who reported doing little to no exercise at the beginning of the study performed worse on the neuropsychological tests at the repeat examination, compared with those who reported regular physical activity at the beginning. Exercise therefore seems to be just as important as diet and cultural pursuits in the overall 'normal' ageing process, as well as for those with dementia-related complications. And, like diet and cultural habits, physical activity can be modified, does not require medication, and can reap benefits that extend to other age-related conditions.

I'm eager to get back to the question of artistic expression, though. In the discussion of the Nun Study in Chapter 1, we saw how the ability to be inventive with language and the possession of a fertile imagination plays a crucial role in how we progress through the ageing process. And the art lab in the Milan nursing home illustrated the role art can play in alleviating or lessening cognitive decline. Playing a musical instrument can protect the brain from developing dementia, even in individuals who pick up an instrument late in life. Reading music, and processing and executing the notes on a musical score, forces us to use numerous different areas and functions of the brain.

In a study recently published in *Frontiers in Human Neuroscience*, participants aged 70 who had completed at least nine years of musical studies were compared with a group of non-musicians of the same age, on a series of neuropsychological tests.[40] The musicians scored higher on working memory, which is a component of short-term memory. Music was also seen to have a protective effect in subjects who learned to play an instrument after age 65, regardless of their level. Modern neuro-imaging techniques have been able to identify changes in the brain structures of musicians. Certain areas are noticeably larger than non-musicians, particularly the thick band of nerve fibres that connects the right and left hemispheres and which transfers visual, auditory and tactile information between the two, coordinating movement and language.

The scene in Paolo Sorrentino's film *Youth* in which Michael Caine (playing a retired conductor and composer) conducts a field of cows becomes all the more poignant when we discover that the character's wife, the soprano for whom the piece was originally written and whose presence is like a constant shadow throughout the film, is in a Venice hospital with dementia. In the film, Sorrentino suggests time is not measured in years but in how much we can remember.[41] We can be as active and creative on the cusp of our eighties as we can be passive and uninspired in our twenties.

For an example of old age at its most venerable, without so much as a hint of dementia, we need look no further than Giuseppe Verdi. The proof is in his musical productivity, which grew more innovative and experimental the older he became. An extreme example is his comic opera *Falstaff*, based on Shakespeare's *Merry Wives of Windsor*, composed when he was approaching 80. Conductor Arturo Toscanini insisted it was the best opera Verdi had ever written.

It's also remarkable how dementia – through the Shakespearean character most associated with it, King Lear – haunted Giuseppe Verdi for most of his career.

In 1850, when he was still in his thirties, he began to (unsuccessfully) commission librettos which could be set to music for an opera of the Shakespeare's tragedy, but he was always discouraged by the librettists, who found it impossible to fit the complex structure of the text into a plausible libretto. It wasn't until he was almost at the end of his life that Verdi finally found one who was willing and able – Arrigo Boito, who had written Verdi's previous two operas.

But Verdi felt 'too old' by then. And what a pity. One can only imagine the powerful score he might have come up with to convey the storm raging at the heart of the tragedy around King Lear.

CHAPTER FOUR

—◆—

Active ageing

If we could give every individual the right amount of
nourishment and exercise, not too little and not too
much, we would have found the safest way to health.

Hippocrates

I like to start my day with a walk. If it's sunny, I go through
the park to stroll in the shade of the plum and lime trees,
and past the ruins of an ancient Roman amphitheatre. I
stop to admire the age-old red bricks of the church of
San Lorenzo. On the days when I leave a bit later in the
morning, the lawn behind the apse will already be dotted
with students, stretched out on the grass with books.
When I get to the university, I usually take a shortcut
through the Guastalla Gardens, the oldest park in Milan,

where the pond is a deep blue in winter and shimmering green in summer, and when the sun hits the neoclassical temple there full on, it turns the columns gold. I come out onto Via della Commenda, walk past my son's old school (whose time there seemed to go by ever faster – his secondary years passing in a flash) and then reach the Policlinico hospital.

There, I weave my way through the familiar pavilions and enter the Polifit gym. Located on the second floor of an impressive fifteenth-century building (the hospital's original building), it has huge windows that look out over the site of the new hospital (established in the late 1800s) and the red-brick San Giuseppe church.

It always fills me with joy when I go into the gym and see so many older adults – several in advanced old age – exercising together. It makes me happy to think that the classes set up for people who prefer to avoid expensive health clubs have not only brought them physical benefits but also lifted their mood thanks to the many new relationships forged. No matter their caregiving commitments – to husbands, wives or grandchildren – these individuals are always on time for their sessions with the much-loved Dr Giuseppina Bernardelli, which is just as well as she doesn't admit latecomers! The exercises vary from week to week and typically feature moderately energetic aerobics (like walking on a treadmill) to tone muscles, improve breathing and strengthen the heart; resistance exercises (using sticks, weights and bands) to build specific muscle groups; balance exercises to help maintain posture, regulate movement and reduce the risk of falls; and exercises for the spine, which aim to prevent the inevitable backache we all have to deal with in old age.

After the first full cycle of exercise classes, consisting of a forty-five-minute session a week for six months, the participants were very happy with the small improvements they had made. One woman, for example, reported being able to put her tights on without any help for the first time in years.

These exercise classes were made possible by funding from the Italian Ministry of Health. Our objective had been to devise an exercise programme specifically geared to elderly citizens, and to make it available in a safe environment in the local community. Programmes like this are referred to internationally as Adapted Physical Activity (APA), a discipline founded in 1973 by Clermont Simard, who was for many years Professor Emeritus at Laval University in Quebec City, Canada. APA comprises a series of non-rehabilitative exercises adapted to the diminished functional capacities of patients with chronic conditions (such as arthritis of the spine, hips or knees). It aims to promote lifestyle changes that will prevent fragility and boost independent living. And it fits well with the philosophy at the heart of geriatric medicine, as it requires a multidisciplinary team of doctors, physiotherapists, qualified nutritionists and sports scientists. When it was first introduced, the elderly population of Quebec experienced undeniable benefits, and all reported clear improvements in their quality of life.

Jean Claude de Potter imported the Canadian pilot experience to Europe in 1979. He brought it to the attention of UNESCO and to sports ministers across the continent, highlighting the preventative benefits to society at large. This fuelled the beginnings of a new culture in which physical exercise was reintroduced as an essential and integral part of everyday life for people with severe mobility issues, allowing everyone the chance to reap its benefits.

Following his lead, the World Health Organization (WHO) more recently coined the term 'active ageing', which it defines as 'the process of optimizing opportunities for health, participation and security in order to enhance quality of life as people age'.[42] It is worth noting that in the WHO definition, 'health' refers to physical, mental and social well-being. We've touched on this multi-faceted definition several times already, but Western culture seems reluctant or at least slow to recognize the importance of the mental and social aspects of personal well-being. This is despite scientific research – like the Nun Study we discussed in Chapter 1 – which has consistently shown how education, imagination and richness of expression are key to keeping the brain youthful and to ageing well.

These three realms – physical well-being, mental health and being part of a cohesive community – are essential at any stage of life, but even more so in our latter years. APA programmes like ours have provided tangible proof, as have the plethora of scientific studies on the subject, that regular exercise can reduce the risk of multiple age-related disorders and slow down the onset of chronic illnesses. The official advice, which should be common sense by now, is that physical activity is recommended at all ages. What's really interesting, though, is evidence from rigorous international research, involving sizeable subject groups, that shows particularly marked benefits in the elderly – especially those in the at-risk category. The Lifestyle Interventions and Independence for Elders (Life) Study is one source of such evidence.[43] It recruited 1,635 sedentary men and women, aged between 70 and 89, with physical limitations but who were nevertheless able to walk short distances without the use of a cane or the help of another person. The participants were selected from eight clinical sites located in urban, suburban and rural areas in the United States (University of Florida, Northwestern University, Pennington Biomedical Research Center, University of Pittsburgh, Stanford University, Tufts University, Wake Forest University and Yale University) and were followed for an average of 2.7 years. Random sampling was then used to split the subjects into two groups: 818 individuals embarked on a moderate-to-intense programme of physical activity, twice a week in the university gym and 3–4 times at home, and the remaining 817 were assigned to an educational programme involving seminars on ageing-related issues as well as a regimen of simple stretching exercises.

After an initial period of observation and half-yearly assessments, the Life Study revealed a significant reduction in major motor disability (the inability to walk 400 metres in less than 15 minutes) and in the combined outcome of major motor disability or death (patients who had a major motor disability and later died) in the patients who were enrolled in the physical activity programme, compared with those who followed only the educational intervention. The greatest benefits were observed in subjects who had lower baseline scores assessing lower-limb functionality. This group made up almost half of the participants.

Despite these results, we know that physical activity alone is not enough to maintain good health; it must be accompanied by a healthy diet – another idea we touched on earlier and will return to later in our conversation on centenarians. For now I'd like to reiterate how important dietary habits are to the ageing process. Since our sense of taste alters with age, many people also lose their appetite as they get older. Dental problems can occur, along with the risk of malnutrition – which should not be confused with being skinny, as there is also a condition in which both sarcopenia (loss of muscle mass) and obesity are present (sarcopenic obesity, in which individuals have disproportionately poor muscle strength).

The term sarcopenia (from the Greek *sarx* [flesh] and *penia* [poverty]) was first coined in 1989 by Irwin Rosenberg, a professor at Tufts University.[44] He used it to describe the loss of body mass and the resultant loss of strength affecting older adults. It is not a disease, just a part of the ageing process. But many studies report a link between sarcopenia and an increased likelihood of falls, disability and death.[45] It is a factor in symptoms like frailty, inactivity and weight loss in the elderly, and also the tendency to walk very slowly and tire easily. Although typical of old age, symptoms are heightened in patients confined to bed, those leading a sedentary lifestyle and those being treated with certain types of drugs (diuretics, for example). Nutrition is also a factor, because of the increased need for protein as we get older. The international PROT-AGE Study Group recommends a specific daily protein intake (dependent on body weight) if malnutrition is to be avoided, which should be raised for individuals with acute diseases, and then again for patients defined as malnourished.[46]

Foods to be included in a healthy diet are fish, white meat and olive oil, as well as plenty of fibre-rich fruit and vegetables and preferably wholewheat grains. Foods to avoid or limit are precooked foods that contain AGEs – not only because of their reported link to an increased risk of developing Alzheimer's, but also because of their typically high glycaemic indexes, which result in a rapid increase in blood sugar levels – something we should try to avoid.

Mindful that the human body is 60 per cent water, we should also aim to drink at least one and a half litres of fluids a day, because dehydration can be a serious problem. Thirst diminishes with age, so we tend to drink less as we get older, making the risk of acute dehydration during the hotter months even greater. By drinking more, this can be avoided easily.

I once read in an interview with Pope Francis that his favourite film is *Babette's Feast*.[47] Directed by Gabriel Axel, it is based on the short story of the same name by Danish author Karen Blixen and won an Oscar in 1988 for Best Foreign Language Film.

Babette is a Parisian chef who flees the Paris Commune in the late nineteenth century and seeks refuge in a frugal Christian community in Denmark, where she becomes housekeeper to two pious Protestant spinsters devoted to doing good works. The images in the final scenes have always captivated me. Babette, having won the lottery, has spent all her winnings on a lavish feast for her hosts and their friends. To the increasing astonishment of the guests, dish upon dish of sumptuous food is brought in, all of which is washed down with the finest of French wines. The sensuality of the scene is made even more vivid by the muted colours, the candlelight, the white Flanders linens, and the tantalising shots of each of the culinary masterpieces served. You can almost taste the food.

Babette's remarkable banquet transforms the mood of the staid guests, helping them to overcome their disagreements and come together to dance in the moonlight after the meal, before they all head off to their respective homes. And it reminds me of something the son of an elderly couple I treat told me a few years ago. He would go to see them every weekend with bags of groceries from a nearby supermarket. He'd fill the fridge with fruit and vegetables, yoghurt, cheese and fresh fish, and the freezer with various different types of meat. Yet on every visit, he'd find the fridge almost exactly as he'd left it. Such a situation is fairly common; I expect it's happened to most people who have an elderly relative in the family. Is it just loss of appetite? Or maybe uncomfortable dentures that make cooking and eating a healthy meal less appealing?

The story for this man didn't end there, though. 'If I invite them over to eat with their grandchildren at our house, rather than buying them groceries, they will eat happily and heartily.' This is something else I'm sure most of us have witnessed. After all, humans are social and cultural animals by definition, so it's only natural that we enjoy eating together with family and friends – and this anthropological trait certainly doesn't change as we get older. Cicero reminds us of this in *De Senectute* (*On Old Age*): 'Nor, indeed, did I measure my delight in these social gatherings more by the physical pleasure than by the pleasure of meeting and conversing with my friends.'[48]

Having lunch with family, and the pleasure that comes with being part of the lives of our children and grandchildren, has benefits for elderly people that go beyond those of a healthy meal. Eating together can be considered, to all intents and purposes, an elixir of long life.

And it's not just eating together. The Berlin Aging Study followed more than 500 people over a period of twenty years, from 1989 to 2009.[49] The participants were all aged over 70. From the data obtained, the researchers saw a marked improvement in the health of active grandparents, and their risk of dying was one-third lower than that of grandparents who did not provide any childcare. Half of the grandparents who cared for grandchildren were still alive ten years after the initial interview, while half of the participants who did not help others died within five years. Such caregiving is useful for society, and also has benefits for an individual's health and longevity.

———— ◆ ————

I don't have to look too far to see how an active lifestyle can have a positive effect on ageing. An elderly cousin of mine – a widow, now in her eighties – continued to travel on her own after losing her husband, and only stopped recently (after trips to the United States and Patagonia). When the Smart car was introduced in Italy, she was one of the first to buy one. She took an IT course run by the City Council of Milan, regularly uses a computer and smartphone, and keeps on top of new developments in a world in which technology evolves at breakneck speed. She buys a theatre season ticket every year, has friends of all ages, and shows so much enthusiasm for life that you can feel it by just standing next to her. When she was in hospital a few years ago, she bolstered the spirits of people half her age. She comforted them and made herself useful in any way she could. The doctors and nurses on the ward still remember her indomitable spirit. And if I ever arrived at the hospital after visiting hours on my way home from work, I'd find her in bed tucked up under the covers, peacefully asleep – a rare phenomenon in a hospital, especially in a geriatric ward.

Elderly patients typically report having problems sleeping, and too often this can lead to an over-reliance on sedatives. Scientific literature is full of examples of the harmful effects on the brain of long-term use of these medications, their side effects often increasing the risk of falls, fractures and car accidents, not to mention the onset of dementia.

Sleep is an important activity in life. While it can cause problems if we get too much of it – with attention, learning and memory, as well as severe aches and pains – generally, the elderly have the opposite problem. Animals with complex brains all seem to need sleep, and a recent study in mice, published in *Science* and funded by the National Institutes of Health, provides a convincing explanation of why.[50] The researchers concluded that synapses (the connections between neurons) grow during waking hours and shrink during sleep, and suggested that this is to protect the brain from overload and allow the mice – and this is probably true for us too – to wake up 'revived' after sleep and ready to start afresh.

Chiara Cirelli and Giulio Tononi at the University of Wisconsin – using sophisticated 3D electron microscope technology to measure these synapses – confirmed these results. They assessed a total of 7,000 synapses in the sensory and motor cortexes of the brains of mice, before and after sleep. They found that the synapses grew stronger and larger at night, when the mice woke up and scuttled around the lab. By interacting with their environment, the animals were learning new things without even realizing it. After six to eight hours' sleep during the day, the same synapses were found to shrink by 18 per cent – the brain had basically 'restored' itself, ready for another night's activity.

The essential thing is not just sleep itself but the brain's ability to 'refresh' after stimulation – and the quality of that stimulation is equally important. As I explained in Chapter 3, recent studies have shown that reading music and playing a musical instrument, even if not until later in life, can prolong cognitive function. The ageing brain is malleable; it is always capable of learning new things, and playing a musical instrument is excellent training. We also know that *listening* to music has health benefits, in addition to offering an escape from the trials of everyday life. Listening to a favourite tune about half an hour before bed can actually make it easier to fall asleep, as well as to stay asleep for the whole night, thereby eliminating the need for sleeping pills. Intense intellectual activity or strenuous physical exercise, on the other hand, are not recommended last thing at night – with the notable exception of love-making, which has been proven to make us live longer.

Interestingly, there is an enduring myth that elderly couples become asexual with advancing age. I remember an 80-year-old woman from a Sicilian village who asked to speak to me in private one day. She had a very discreet air and was dressed elegantly, although in an old-fashioned way. Her husband had been called into the clinic for an ECG and their son had brought them in. When we were alone, in a shaky voice and almost in tears, she told me (and I quote), 'I don't know what to do. My husband wakes me at night to partake of my flesh…'

After this plea for help, I was sitting in a meeting with some postgrad students one day when I realized that younger generations in particular are reluctant to ask about the sex lives of elderly patients when preparing their case notes. In the special ageing issue of *The Gerontologist*, Gayle Doll from the Kansas State University Center on Aging presented the results of a study that knocked the bottom out of the widespread belief that old people are uninterested in such things.[51] Doll had sat with five couples, who had all been together for more than fifty years, to watch a selection of romance films – some of which featured explicit sex scenes. The films they saw included *Hope Springs*, portraying a couple trying to rekindle the flame; *45 Years*, in which the central couple's relationship starts to crack when the dead body of the husband's former lover turns up unexpectedly; *Still Mine*, in which a man is willing to do anything to build a new house for himself and his ailing wife when their existing home is no longer suitable; *Cloudburst*, about an older lesbian couple who embark on a *Thelma and Louise*-style road trip across America and Canada to get married; and *Cloud 9*, a film dealing with sexuality in later life. In the discussions afterwards, the couples confirmed, each in their own way, that intimacy and feelings were important in their relationship, even if the passion and lust of youth were long gone.

Love, in all its forms, improves health and well-being and makes us live longer – statistics collated in geriatric research prove it. For several years, scientists have been looking not just at biomedical factors but also at familial and social factors, such as whether someone is single or part of a couple, to determine the effect of each on older adults.

Just a few days before I started writing this book, a 65-year-old patient came to see me. I'd treated her about two years previously for reactive depression after an acrimonious divorce from her celebrity husband, who had left her for a student. On top of her own pain and that of her daughter, she had to cope with the professional and personal humiliation of being dragged through the mud in the press. With the help of a psychoanalyst, she had managed to get her life back on track. As she told me that her daughter had recently married, I noticed how radiant she looked. I wasn't surprised when she announced she herself was about to remarry.

Going back to *The Gerontologist*, that February 2017 edition featured an article by researcher Ruth Ray Karpen from Wayne State University, Detroit, in which she recounts her personal experience as a baby boomer and shares an interesting insight that emerged in her research.[52] She had discovered that women are more positive about retirement than men and more curious about what lies ahead. They are not scared to end stale relationships and start over, and some even marry at the age of 63 – as the author herself did, after a chance meeting with a long-lost friend. What she describes in the article is a revival of her soul, just like the one I witnessed in my patient. This kind of physical and spiritual renewal can help patients get the most out of the years they have left.

———◆———

A geriatrician's task is to offer a more in-depth treatment programme than would be recommended for a younger adult. Elderly people often take more medications, prescribed by specialists who may not have consulted one another, and they often suffer from multiple conditions. Caring for elderly patients, therefore, is not only more complicated, but also more time-consuming, as it takes longer to explain the various diagnoses and employ the necessary empathy to understand what the patient wants, while still assessing the risks and benefits of any new treatments. Evidence-based medicine (EBM) guidelines drawn up by scientific companies to help practitioners make the best decisions about the care of an individual patient, based on up-to-date, solid scientific evidence obtained in controlled clinical studies, are often aimed towards younger adults with only one illness. As a result, these guidelines can be contradictory and too difficult to apply to elderly people coping with age-related changes, who are often affected by multiple conditions and taking numerous medications.

Consequently, geriatricians know when to reach out to experts in other disciplines to find creative solutions that are tailored to the particular needs of the elderly. A geriatrician will find out what the patient hopes to achieve and what is most important to them, and will make sure the resulting pharmacological intervention reflects this. In practice, this means making tangible improvements to a patient's quality of life so that our patients feel healthier, remain independent for longer, and are less likely to need hospital services or be admitted to a nursing home.

In today's world of ever-more-crowded hospitals, healthcare services need to focus more on illness prevention and on keeping patients in their own homes. Young geriatricians need to be trained to pick up on risk factors and understand the constraints imposed by ageing, physiology, disease, drug treatments and by a patient's economic and social status. I try to impress on my students the importance of such considerations, encouraging them to work with patients and their GPs to gain the necessary perspective from which to spot any health risks before they become a problem.

Patients who come in for a geriatric exam are often frustrated; they not only struggle to understand the various different medicines they've been prescribed, but often feel no one really listens to them, their problems routinely shrugged off with a 'What do you expect at your age?' There is clearly a stigma in the Western world around advanced age, propagated not only by society but by the patients themselves, and dismissive responses from health workers only make this worse.

Working towards overcoming preconceptions and trying out new treatments for age-related conditions is beneficial, not just to the patients – who go home with fewer drugs and a newfound enthusiasm for an improved lifestyle – but also for their geriatricians, who, according to research, report higher levels of job satisfaction on average than other specialists.

The technological revolution we are currently living in has spawned a wealth of excellent alternatives to conventional therapies. New technology has transformed every area of our daily lives, from how we communicate and source information and news, to how we do the simplest of things: meeting someone new, booking a table at a restaurant or checking train times, for example. It can also help us to overcome the everyday difficulties we encounter when we are old and alone.

I represent the foundation hospital of Milan as a geriatrician in the MoveCare (Multiple-actOrs Virtual Empathic CARegiver for the Elder) project, funded under the EU's Horizon 2020 programme and coordinated by the University of Milan, in particular by Professor Alberto Borghese.[53] MoveCare brings together international experts from both academia and industry, the common goal being to prolong the independence of the elderly. This includes those living alone and those living in sheltered housing, all of whom, MoveCare believes, require a simplified lifestyle in order to go about their everyday activities – as well as personal alarms able to detect any health changes.

In the MoveCare platform, individuals are assisted by a robotic companion (called Giraff), which interacts with various pieces of hardware and software in the home, known collectively as a domotics system. This consists of 'smart' (technologically intelligent) objects, which may include, for example, mugs with sensors that detect a drop in physical strength, ergonomically designed pens that are easy to hold, and motion-sensor pads worn in shoes to pick up any changes in gait or balance.

The platform also comprises a virtual community, which can be either local or more wide-scale (depending on user preference) and which can be used to play games. The community also provides support, offering access to a help centre that uses artificial intelligence to monitor activity and which can provide guidance when required. MoveCare has a modular design, which means it can be tailored to meet the needs of each individual, letting a person choose the voice and appearance of their robotic companion, for example. Would they prefer the screen to display a familiar face, or that of a favourite actor? And whose voice would they like to wake them up in the morning? At this early stage of our research, we are working with a group of GPs, geriatricians and elderly adults to determine what exactly we want the domestic robot to do. Some potential users would like it to remind them how to make a family recipe or when to take their pills, or enable them to chat remotely with their children and grandchildren, or even to clean the house, while others dream of having a personal trainer (at least, the ones who like exercise do), someone to play cards or chess with, or a way of calling their GP about health issues.

Roberto Cingolani, a leading Italian physicist, robotics expert and director of the Italian Institute of Technology, ends his book on nanotechnologies – *Il mondo è piccolo come un'arancia (The World Is as Small as an Orange)* – with a chapter looking at nanomedicine and bio-inspired technologies. He describes hearing aids that use hair cells to detect sound-generated vibrations in the inner ear; tactile prosthetics, like artificial limbs with synthetic skin; and intelligent robots to assist in the rehabilitation of neurological patients. A few pages later, the attention switches from rigorous academic observation to futuristic fiction, when those same discoveries become the protagonists of a short sci-fi tale set in a space-age world that is perhaps not as far away as we think.

The passage narrates the life of an elderly man living in this near-future world:

23 October 2064, Rome, megalopolis. An ordinary day in the life of an ordinary man. His first thought when he wakes up is – eighty-seven years old. He imagines the numbers, big and black – 8, 7. His memory presents him with the following information: eighty-seven is the sum of the squares of the first four prime numbers – two, three, five and seven. As he thinks about this, he counts, and as he counts, hands connect haematic sensors to his arms and respiratory sensors to his mouth. He hears the beep, beep, beep of the neural decoder and before it gets to 'seven', in the quiet and almost imperceptible hum of the night-time silence still pervading the room, the wall-display switches on, bathing the bed in light. 'Good morning,' he says to the smiling doctor on the screen. [...] He is putting on his dressing gown when there's a knock at the door. 'Come in.' Jeeves-3 opens the door and glides in. 'Good morning. My calendar informs me that today is your birthday. Felicitations. I hope you are hungry because Chell-2 has made your favourite, a forest fruits tart. Your blood sugar level is excellent at the moment so you can have as much as you wish. I will look the other way for once. The maximum temperature today will be 27°C and we will have a low of 20°C . May I recommend a lightweight jacket? There is only a 14.7% chance of rain, so an umbrella won't be necessary.' As Jeeves-3 speaks, he distracts him from the needle prick in his arm.

I wonder if Isaac Asimov, who first popularized robotics in science fiction with his 1942 short story 'Runaround', would have believed that the world he foreshadowed might become reality so quickly. Perhaps the fact that I'm a lifelong fan of the *Star Wars* saga by George Lucas explains why I find the idea of growing old

with a trusty protocol droid like C-3PO and his insepa-
rable sidekick R2-D2 (called C1-P8 in the Italian version
of the original trilogy) by my side really rather appealing.

CHAPTER FIVE

———◆———

Learning from the Centenarians

It's all relative. Take a supercentenarian who breaks
a mirror: he'll be happy to know he has seven more
years of bad luck.

Albert Einstein

I took the first boat from Portofino to San Fruttuoso one
day last summer. I wanted to visit the abbey before the
hordes of beach-goers arrived. The tiny bay was as
beautiful as ever in the morning breeze, and the resto-
ration work done by Italy's National Trust made my visit
to the abbey even more special. The sun glinted on the
small slate cupola, the mullioned windows in the tower
seemed untouched by time, and the maritime pines around

the building leaned benevolently forward, casting long, luxuriant shadows as their reflections swayed in the sea.

I went into the chapel alone, and as I wandered through it, I came across the icon of St Charalampias, martyred in AD 202 at the age of 113. He was celebrated for his healing ability, but he also had the power to make land fertile and many farmers still venerate him today for this reason. Although I have been studying centenarians for many years now, I had never heard about this saint. He would have made an amazing case study.

———— ◆ ————

When the protagonist of *Lost Horizon*, on his quest to find the legendary Shangri-La, finally sees the monastery, his sense of wonder is clear:

> *It was, indeed, a strange and almost incredible sight. A group of coloured pavilions clung to the mountainside with none of the grim deliberation of a Rhineland castle, but rather with the chance delicacy of flower-petals impaled upon a crag. It was superb and exquisite. An austere emotion carried the eye upward from milk-blue roofs to the grey rock bastion above, tremendous as the Wetterhorn above Grindelwald. Beyond that, in a dazzling pyramid, soared the snow-slopes of Karakal.*

Our own journey, which began with this legendary concept has treated us to many breathtaking views, which have been sublime in their ability to elicit both admiration and fear. We have philosophized with Schopenhauer; visited a monastery and heard the memories of young religious sisters; witnessed the birth of David Copperfield and learned that we are born old, despite the best efforts of the good doctor; we flew over the battlefields of World War Two, which brought famine upon entire villages; peeked into the lab of Professor Alzheimer, saw him hunched at his desk examining samples from his patient's brain; then we raised our eyes and realized how the whole world, its nature and its art, is the most powerful antidote that we have to ageing. We even encountered mini-androids who are waiting to take care of us in the future.

But there can be no marvel great enough to equal the sense of wonderment I feel every day working with centenarians and supercentenarians. It is *they* who are our Shangri-La – and we don't have to fly to the Himalayas to find them.

When I returned to Milan after my visit to the San Fruttuoso abbey, I received a phone call from a journalist in Florence I had befriended some time ago. His father, also from Milan, had been the first centenarian recruited to my study on ultra-longevity. The phone call was to tell me that his father was looking forward to my next visit, and this got me thinking. The first time I met him, he had looked at least ten years younger than his biological age. He was a witty, intelligent man, who had survived the Spanish flu pandemic that killed millions of people between 1918 and 1920 – more than World War One. After his wife's death, he had chosen to live alone, having a daughter nearby who was able to watch over him. I have been invited to his birthday celebrations every year since we met, always held in a nearby restaurant. We would all order a light meal and have only a small sliver of cake. He always ordered the biggest portion of fried fish.

I often think about him when I find myself pondering, after all these years, what made me want to study centenarians. All I know is that, as a geriatrician, I have always been interested in population demographics and the way they are constantly changing, and I am fascinated by what the future might hold – especially given the higher-than-ever-before percentage of elderly citizens in the world and the potential consequences of this for us all. If life expectancies continue to increase at the same rate, this percentage will accelerate during this millennium, producing more and more centenarians.

Statistics have been estimating this for years, predicting a veritable boom by 2050. The world's centenarian population is expected to grow eightfold in the next thirty-five years, rising from the current 451,000 to almost 3.7 million. And almost half of these people will live in five countries – Japan and Italy, which are ageing the fastest, along with the United States, China and India.[54]

Let's look back at the figures I gave you in the first chapter, but from a different perspective. In 1921 there were only 49 centenarians in Italy out of a total population of almost 40 million, and by 1981 there were 1,304 out of 56.5 million inhabitants. Just twelve years later, the national census revealed that the total population in Italy was about the same, while the Italian Multicentric Study on Centenarians reported that around 6,000 of these were centenarians. On 1 January 2015, ISTAT published figures showing that roughly one in every 3,184 inhabitants of Italy was a centenarian. The number of people aged over 105 was 878 (one in every 69,243 inhabitants), whereas the number of people living longer than 110 years was much smaller, only 17 (one in every 3,576,212). Population data published for Milan in 1992 showed 325 people born between 1887 and 1892, relative to a total of 413,335. Figures from 2016 highlight that there are

670 people aged 100 or over in Milan, out of a total city population of 1,345,851.

Almost every other country in the Western world will have similar figures, confirming the obvious truth – that the number of centenarians is set to keep rising. The UK, for example, recently reached a record number of 14,570 centenarians – four times as many as twenty years before, according to the ONS in 2015. In the last decade alone, the number rose by 65 per cent. And this increase was mirrored in citizens living more than 105 years – who numbered 850 in 2015, compared with only 130 in 1985. The total number of centenarians increased in every part of the UK: 71 per cent in England and Wales, 77 per cent in Scotland and 80 per cent in Northern Ireland. The majority are women. A video of centenarian and pianist Gladys Hooper at the age of 112, telling her son how she saw the first German airship shot down in England on the night of 3 September 1916 (an action which won pilot William Leefe Robinson, No. 39 Home Defence Squadron, the Victoria Cross), has been watched more than 850,000 times.

Statistics aside, the prevalence of centenarians reflects an important natural pattern that can help us to understand the key mechanisms behind homeostasis (the body's natural tendency to find a stable equilibrium in its physiological processes) and the psyches of the long-lived. My research team recently shared with the scientific community the idea of forming a worldwide consortium to study those with extraordinary longevity and combine our relative skills and knowledge across a much wider and more significant subject sample.[55]

One of the first female centenarians I studied in Milan was a 104-year-old woman called Agata. She lived in a detached home with her daughter-in-law, the widow of her only son, as well as a granddaughter who was married with a three-year-old son. In an early interview filmed for television, the woman changed her dress and accessories twice and carefully brushed her snowy white hair before leaving the bedroom. When I was in the room alone with her later, I politely commented on how affectionate her daughter-in-law was with her; I'd noticed she'd been very attentive and helpful. The old woman looked at me and replied sarcastically, 'Do you want to know why?' She then made a sweeping gesture towards the ceiling and the exquisite wood moulding, rings glittering on her fingers, and added, 'Because all this is mine.'

I was reminded of Agata and her manner a few years later, in 2006, when 98-year-old director Manoel de Oliveira presented *Belle Toujours*, sequel to the cult film *Belle de Jour* by Luis Buñuel, in Venice. I was in the small theatre where the film was showing and happened to spot Mario Monicelli, the famous Italian director and screenwriter. A journalist saw him, too, and went over to ask if Monicelli found it frustrating that de Oliveira was still so productive even though he was so much older. Monicelli said no, but that he did hope it meant de Oliveira would die soon, making him the world's oldest living director.

High self-esteem and a positive attitude are undoubtedly helpful personality traits to have when it comes to ageing well. I recognized them in Emma

Figure 3. Emma Morano, 117 years old, with Dr. Federica di Berardino and Sarah Mastrillo, a student of Milan University. (© Daniela Mari)

Morano, until recently the world's oldest living person. I was lucky enough to meet her, and she took part in our centenarian study three years in a row. Having lived for 117 years and 137 days by the time of her death on 15 April 2017, she was the oldest living person in Italy from 2 April 2013 and the oldest living person in the world from 13 May 2016.

She lived alone in Verbania, separated from her husband, on the west side of Lake Maggiore. Initially, she worked in a jute factory, then later, until her retirement at the age of 75, as a cook in a boarding school. It wasn't an easy life, but she remained upbeat throughout. During our last interview together, which was on her birthday, she told me for the first time that she enjoyed a glass of homemade liqueur every night. She struggled to stifle a giggle as she said this.

How did Emma Morano reach such a momentous age? Perhaps a new concept, recently introduced into longevity studies, holds the answer: the idea of re-silience.[56] Resilience can be described as a dynamic process whereby an individual successfully employs effective strategies when faced with adversity. It can also be defined as the ability to bounce back from negative emotional experiences and adapt successfully to the changing demands that stressful events place on us.

Emma Morano and Mario Monicelli – and of course Agata, with her carefully brushed hair and sparkly rings – had a healthy dose of it. And this unyielding quality they shared seemed to make them impervious to the passing of time, like the mullioned windows in the San Fruttuoso abbey which are still there, fragile yet steadfast, holding up in the face of persistent rain and saltwater spray.

A study was carried out on the subject of resilience in 2016.[57] A total of 119 individuals aged between 95 and 107 were recruited from the New York City voters' list. They were asked questions about their health, how independent they were in their daily activities, and if they had any children or social support services around them. Centenarians accounted for 52 per cent of the sample. A large number of the participants lived at home, and 75 per cent were widows or widowers. One-third held a university degree, another third held a college diploma or had attended university, while the final third had completed lower secondary or a few years of high school. Only 3.3 per cent had only been primary school educated. Less than 20 per cent of the sample group were classified as clinically depressed. Despite the reduction in physical function and fewer social resources, the elderly participants mostly had good mental health. This would suggest the same high level of resilience and the ability to adapt to age-related challenges that I saw in Emma Morano. And a large number of the long-living individuals in this study were actively involved in the local community, further highlighting their desire to lead an autonomous life.

Similar studies on centenarians in Europe have revealed how these long-living individuals report their health as being either good or excellent, despite the obvious decline in motor and cognitive function – which would seem to point once again to the idea of resilience. People who live this long develop a psychological strategy that helps them to adapt to or cope with biological decline.

Rudolf Arnheim, the art theorist and perceptual psychologist of the Gestalt school who lived until he was 103, proposed a diagram depicting an arch overlaid by an ever-ascending flight of steps.

The arch represents biological lifespan. It starts with a period of growth, followed by one of maintenance, then one of progressive decline. The staircase, on the other hand, represents our ability to be creative, to think and to learn – which is infinite, extending beyond the confines of our years and our lifetime.[58]

Once again on this journey of ours, we must put our faith in a master from the past. Centuries ago, Cicero said:

[...] the most suitable defences of old age are the principles and practice of the virtues, which, if cultivated at every age, bring forth wonderful fruits at the close of a long life lived to the full, not only because they never fail us even at the very end of our existence — although this is the most important thing — but also because it is most delightful to have the consciousness of a life well spent and the memory of many deeds worthily performed.

This consciousness and memory are why centenarians represent the best possible example of human longevity for gerontological research. Through them, we can identify the biomarkers of healthy ageing as well as any factors which may help to ward off the main age-related pathologies, both mental and physical.

In a recent interview, Professor Claudio Franceschi from Bologna University, who coordinated many of Italy's centenarian studies, emphasized how this demographic represents either a rare group (those aged over 100) or a very rare one (those aged more than 105 years).[59] He explained that most centenarians and supercentenarians reach this gold standard because they have managed to escape altogether – or at least put off for longer – the onset of the most common age-related disorders, and they remain physically and cognitively healthy. Italian research findings and literature on the subject reflect this: the secret of centenarians' healthy ageing and longevity is having escaped the major diseases. But while it may seem obvious that to live a long life it is important to be healthy, in practice it is not so easy to achieve.

There are several factors which need to be considered in gerontology. Ageing is associated with chronic or mild inflammation, for example, something which Franceschi, as we learned in Chapter 1, refers to as inflammaging. Such inflammation is the major cause of many – if not all – age-related disorders. Recent studies have shown that centenarians have a balance of pro-inflammatory and anti-inflammatory networks, and it is this that could be the biological key to their extraordinary health.[60]

In addition to the longstanding studies that have illuminated our understanding of ageing so far, some newer studies take a different approach. Researchers at the Harvard Stem Cell Institute, Stanford University and California University, for example, carried out rodent studies, the results of which were published in *Science* and *Nature Medicine*.[61] They highlighted how the blood of young mice could counter or even reverse the effects of ageing when it was shared with older animals. The technique they used involves joining together the circulatory systems of two genetically identical animals of different ages, so that they share the same blood. After four weeks of this treatment, the researchers saw marked improvements in muscle and cognitive function in the older mouse, and that the stem cells in these areas had begun to produce neurons and muscle tissue.

The scientists leading these studies also showed that similar effects could be obtained through straightforward blood transfusions, and with injections of specific proteins which are present in large quantities in the circulatory systems of young mice. The older animals given the transfusions showed marked improvements in motor and cognitive function, outperforming animals of the same age on all physical and memory tests (such as the classic maze-navigation test, for example). Another significant finding was that the benefits were not merely transitory, but instead lasted for several weeks after the injection. Since the data obtained is still experimental, more detailed research is required before such anti-ageing techniques can be extended to humans, but these findings still open up exciting new possibilities and definitely take us somewhere we have never been before.

And there's more. Centenarians and semi-super-centenarians (those living for more than 105 years) also make ideal 'super-control' groups in studies exploring the biological aspects and pathogenesis of age-related diseases. As I mentioned earlier, extra-long-lived subjects may have genetic, epigenetic and metabolic risk factors for such conditions (both cardiovascular and neuro-degenerative), but they don't actually develop the diseases, hence why they live longer. This biological predisposition to evade disease suggests that in such individuals the associated risk factors are insufficient and that other factors (such as environment or lifestyle) must be present, too, in order for diseases to develop. (Just like artists, diseases need special conditions to express their inner creativity.) A number of potentially harmful genes may be 'suppressed' in people who live longer. So far, studies suggest that such longevity is the result of the interaction of genetics (which accounts for roughly 25 per cent of variation), epigenetics, lifestyle and random events.

Mindful of such facts, European geneticists recently performed a genome-wide association analysis of a sample group of over 180,000 individuals.[62] The study analysed their genes to identify potential variations between subjects which may in turn be linked to ageing traits (like disease, for example). By analysing cell DNA using modern techniques, it is possible to identify variations in a single nucleotide – SNPs (pronounced *snips*, and standing for 'single nucleotide polymorphisms') – at any point along the DNA sequence.

A new genetic marker was observed on a particular chromosome in a sub-group of individuals aged between 90 and 105. Those who possessed the marker showed a range of unique traits, such as a lower risk of death from cardiovascular disease, lower risk of coronary disease, lower blood pressure and a lower risk of death in general. Interestingly, the same marker was also identified in a study of 2,178 Han Chinese individuals, as well as among the American centenarians recruited to the New England Centenarian Study.[63] These are only a few of the research areas that recent discoveries and advances in technology have opened up, but they all seem to agree that centenarians may hold the secret to extending and improving lifespan for others.

Another study, carried out recently by Steve Horvath at UCLA, examined the peripheral blood cells of supercentenarians for epigenetic biomarkers of ageing – specifically looking for a group of markers collectively referred to as the 'epigenetic clock'.[64] The results showed clearly that the subjects (all of whom were 105 years or older) were on average 8.7 years younger than their chronological age, and their children also had a lower biological age (5.2 years younger) than an age-matched control group who were not born to long-living parents. On hearing such surprising results, my students are always amazed and tend to ask me the same question, but it is one which, surprisingly, has nothing to do with medicine or biotechnology. Instead, they ask what centenarians eat. However, this isn't as silly as it sounds – well, not entirely – because good nutrition is also crucial to living a long and healthy life.

Generally speaking, centenarians eat in moderation. They drink small amounts of wine at mealtimes and are a normal weight. As the German philosopher Ludwig Feuerbach said, 'Man is what he eats,'[65] and advanced research on centenarians has shown that intestinal flora does, in fact, play an important role in ageing. Indeed, the bacteria living in the human body, the functions of which are closely linked to nutrition, seem to limit the build-up of pathogens and infections, boosting the immune system and triggering systemic metabolic effects. The make-up of our microbiota changes as we grow older, and the ability to preserve 'good' bacteria is directly correlated to the likelihood of ageing well. The biodiversity of intestinal bacteria drops in old age, potentially causing inflammation and imbalances in gut bacteria – which, in turn, can be a precursor to numerous diseases. This would suggest that preserving the biodiversity of bacterial flora through a healthy diet can increase life expectancy.

These results all seem to point to the idea that longevity is linked to keeping cells and organs working efficiently in order to combat the inevitable functional decline that comes with age. Some genes seem to have a role to play in all this, and we know that a single 'favourable' genetic variation can extend life expectancy significantly. Epigenetics is also emerging as one of the key mechanisms affecting lifespan, as exemplified in calorific restriction. As we touched on earlier in this book, it has been shown that eating fewer calories, without malnutrition, is linked to increased life expectancy in both humans and primates, and this can be explained by genes 'switching' to energy-saving mode when food is scarce.

In light of these discoveries, we are faced with an extraordinary truth: longevity can be 'constructed'. Genetic make-up and biological good fortune aside, we ourselves are largely responsible for our fate. This means that we can choose to live in a way that enhances our chances of a long and healthy life, even if we did not inherit genes that might naturally extend our journey towards becoming a centenarian. Data from recent studies shows that human longevity is associated with slower cell growth and metabolism.[66] Genes and gene expression can be 'steered' towards longevity, primarily through diet and general lifestyle.

Looking ahead, I see a future in which mankind will have learned how to live longer, heathier and happier lives. We hold the knowledge in our hands – what happens next depends solely on what we do with it.

Conclusion

> Indeed, the composition of this book has been so
> pleasant to me, that it has not only brushed away all
> the vexations of old age but has made it even easy
> and agreeable. In truth, sufficiently worthy praise
> can never be given to philosophy, whose votaries
> can pass every period of life without annoyance.
>
> Cicero, *De Senectute*

Milan, July 2017. It's been hot and sticky these past few weeks in the city, so I decided to go up to Cervinia for a few days, to seek somewhere cooler to finish this book. The view from my window of what John Ruskin called 'the most noble cliff in Europe'[67] has held me spellbound for many years. I love how it stands firm and constant, yet varies with the passing seasons, the changing light, and from one hour or even one minute to the next. It will be almost transparent at the first light of dawn, radiant as the sun rises, and suffused in a pink glow when it sets, seeming even higher and more invincible in the pewter night sky.

After many a winter and summer spent skiing on the beautiful trails around Mount Cervino, as the Matterhorn is known in Italy, I decided to explore the mountain a bit more this time. I started out on foot from the town and made my way up – not using the cable car – to the Carrel Cross (2920m). This spot commemorates Jean-Antoine Carrel, the renowned mountain guide and leader of the second expedition to conquer the Matterhorn back in 1865. I took the ascent slowly, keeping my eyes down and being careful to scan the ground ahead of me, battling both nerves and fatigue. When I did look up, though, I caught unexpected glimpses of glaciers, waterfalls and the peak itself, and the taste of the rarefied mountain air filled me with joy and made the effort worth it.

Much like climbing a mountain, scientific research requires constant effort and can often feel repetitive and unsatisfying, but when you begin to see the idea that inspired you bear fruit, through significant results corroborated by others, all the problems and disappointments of the past disappear and you feel compelled to go further. Yet a lifetime is not enough to reach the top and find explanations for everything, let alone a phenomenon as complex and multifaceted as ageing. The only thing we can do is continue in our research, and aim not only to help people live longer but also to make sure their old age is as healthy and stimulating as possible.

During my short break, I was excited to read *Zero K*, Don DeLillo's most recent novel, as another great work of his, *Underworld*, has always been one of my favourites. In *Zero K*, he describes magnificently the stark contrast between a father and son. Ross, the father, is never around, more preoccupied with his wealth and seeking immortality for himself and his beloved second wife, Artis, who has multiple sclerosis. He funds a dubious organization offering cryopreservation (which is called, significantly, Convergence) in the hope of finding the key to everlasting life. Jeffrey is his son, who has a very natural, almost profound recollection of his mother's death, which is engraved – seared, if you like – in his mind. He has only his girlfriend in whom to seek comfort. It is this poignant yet peaceful beauty that he recalls throughout the novel.

I'd never felt more human than I did when my mother lay in bed, dying. This was not the frailty of a man who is said to be 'only human,' subject to a weakness or a vulnerability. This was a wave of sadness and loss that made me understand that I was a man expanded by grief. There were memories, everywhere, unsum-moned. There were images, visions, voices and how a woman's last breath gives expression to her son's constrained humanity. Here was the neighbour with the cane, motionless, ever so, in the doorway, and here was my mother, an arm's length away, a touch away, in stillness.

'Everybody wants to own the end of the world,' the novel begins, but octogenarian DeLillo offers no answers and leaves his readers to do their own soul-searching about whether technology, at its most futuristic, can ever conquer the limited lifespan of man. What does seem certain is that research into ageing will rely more and more on new technologies, to understand how to keep body and mind working efficiently and to discover what help can effectively be offered during old age.

DeLillo also mentions Martin Heidegger in the novel. The German philosopher was one of the first to refute the assumption that technology prevails over self. 'Meanwhile man, precisely as the one so threatened, exalts himself to the posture of lord of the earth,' he wrote in 1977. 'In this way the impression comes to prevail that everything man encounters exists only insofar as it is his construct. This illusion gives rise in turn to one final delusion: it seems as though man everywhere and always encounters only himself.'[68]

I can't help thinking how different philanthropic composer Verdi was from DeLillo's fictitious Ross, the billionaire father fit for a *Newsweek* cover. Verdi bequeathed most of his wealth to less fortunate artists, singers and musicians, by founding a retirement home for them in Milan. He himself wished to be laid to rest there on his death, stating, 'Among my works, the one I like best is the Home that I have had built in Milan for accommodating old singers not favoured by fortune, or who, when they were young did not possess the virtue of saving. Poor and dear companions of my life! Believe me, my friend, that Home is truly my most beautiful work.'[69]

Through both his memorable operas and this final gesture, Giuseppe Verdi achieved true immortality, of a kind that embraced the less fortunate in the world for centuries to come. Now more than ever, this generous decision to offer tangible help to the elderly – not only for his peers but also for future generations – feels important. Our modern consumerist society's preoccupation with what's visible, reflected in our relentless quest to appear eternally youthful, has resulted in a reliance on money as the only route to power and success.

As philosopher Stephen Cave suggests in his book *Immortality*, it is only the 'works' we leave behind – the things which outlive us – that attain true immortality. In the same way as a book is in one sense limited by its covers, human life is bound between birth and death. But a book, even when closed, can take its reader to distant places and on fantastic adventures.

Human life should be the same. Imagine the book of your life – don't be afraid of what happens outside the covers, and don't worry about how long it is. The only thing that matters is the story inside.[70]

Acknowledgements

There's a photo that has followed me everywhere, always present on my desk over the years. It is of the first Italian longevity research group, made up of Claudio Franceschi, Paolo Sansoni, Giovannella Baggio, Giovanna de Benedictis and myself: I'd like to thank each of these wonderful people for encouraging me to take my first steps in the world of centenarians. My thanks also go to Daniela Monti, Giuseppe Passarino and my lifelong friends Lella Coppola and Bianca Bottasso. Thank you to all the young people who, with their never-ending enthusiasm, helped us to achieve what we have, including Giulia Ogliari, whose talent has seen her snapped up overseas, and many more whom I'm sorry I can't name individually. A special expression of gratitude goes to Giovanni Vitale, who brought his endocrinological expertise to the geriatric field of study, and to Beatrice Arosio, who manages adeptly and enthusiastically our ageing research lab. I am eternally grateful to the people I work with – Paolo Rossi, Marco Ferretti, Tiziano Lucchi, Giuseppina Schinco, Cristina De Fazio, Simona Ciccone, Sarah Damanti and Giulia Dolci – who care for fragile patients in the face of enormous difficulties, and Maura Marcucci, a talented professional recruited outside of Italy. Thankfully, I was able to endorse the University of Milan's recruitment of Matteo Cesari under the Italian Ministry of Education, University and Research's programme to reverse 'brain drain', and I am happy that everything I have built over the years will be in safe hands, as will geriatric care in Milan. I am also indebted to Angelo Recalcati from the Itinera Alpina antique book shop, for allowing me access to the books and for his learned suggestions. And thank you to my publisher Il Saggiatore, especially Andrea Morstabilini, whose understanding and skill were of inestimable help.

My final thanks go to my son Pier Vittorio, whose belief and optimism dispelled any doubts I had about whether I could write this book.

Bibliography

Arnheim, Rudolf, *New Essays on the Psychology of Art* (Berkeley: University of California Press, 1986)

Asimov, Isaac, *I, Robot* (New York: Gnome Press, 1950)

Cave, Stephen, *Immortality: The Quest to Live Forever and How It Drives Civilization* (New York: Crown Publishers, 2012)

Cicero, Marco Tullio, *Cato Maior De Senectute (Cato the Elder on Old Age)*

Cingolani, Roberto, *Il mondo è piccolo come un'arancia* (Milano: Il Saggiatore, 2014)

Deledda, Grazia, *Sino al confine* (Milano: Treves, 1910)

Dinesen, Izak [Blixen, Karen], *Anecdotes of Destiny* (New York: Random House, 1958)

Einstein, Albert, 'On the Electrodynamics of Moving Bodies', in *The Principle of Relativity* (London: Methuen and Company, 1923)

Feuerbach, Ludwig, *Das Geheimnis des Opfers, oder Der Mensch ist, was er ißt* [*The Mystery of Sacrifice, or, Man is What He Eats*], trans. Cyril Levitt (2007)

Genova, Lisa, *Still Alice* (Bloomington: iUniverse, 2007)

Heidegger, Martin, *The Question Concerning Technology, and Other Essays*, trans. William Lovitt (New York and London: Garland Publishing, Inc., 1977)

Hilton, James, *Lost Horizon* (London: Macmillan, 1933)

Hippocrates, *Opere*, ed. Mario Vegetti (Torino: UTET, 1965)

von Hofmannsthal, Hugo, *Der Rosenkavalier* [libretto], 1911

Leopardi, Giacomo, *Zibaldone: Pensieri di varia filosofia e di bella letteratura*, ed. M. Caesar and F. D'Intino (Firenze: Successori Le Monnier, 1900)

Munro, Alice, *Hateship, Friendship, Courtship, Loveship, Marriage* (Toronto: McClelland & Stewart, 2001)

Treccani Maddalena, Mario Rivardo, *Il disegno e i colori nella clinica dello psicoanalista*, 2 vols. (Milano: Scalpendi, 2012)

Robine Jean-Marie, Bernard Forette, Claudio Franceschi and Michel Allard (eds.), *The Paradoxes of Longevity* (Berlin: Springer-Verlag, 1999)

Ruskin, John, *The Stones of Venice*, 3 vols. (Boston: Dana Estes & Company, 1898)

Russell, Bertrand, *The Conquest of Happiness* (London: George Allen & Unwin Ltd., 1930)

Sacks, Oliver, *The Man Who Mistook His Wife for a Hat, and Other Clinical Tales* (New York: Summit Books, 1985)

Schopenhauer, Arthur, *L'arte di invecchiare ovvero Senilia*, trans. Franco Volpi (Milano: Adelphi, 2006)

Utermohlen, William, et al., *Portraits from the Mind: The Works of William Utermohlen, 1955–2000* (Chicago: Alzheimer's Association, 2008)

Verdi, Giuseppe, *Autobiografia dalle lettere*, ed. Aldo Oberdorfer (Milano: Rizzoli, 1951)

Wilde, Oscar, *The Picture of Dorian Gray* (London: Ward, Lock & Company, 1891)

World Health Organization, *Active Ageing: A policy framework* (Geneva: World Health Organization Publications, 2002)

Notes

INTRODUCTION

1   In *Lost Horizon*, the Shangri-La guide Chang explains the community's secret: 'If I were to put it into a very few words, my dear sir, I should say that our prevalent belief is in moderation. We inculcate the virtue of avoiding excesses of all kinds – even including, if you will pardon the paradox, excess of virtue itself [...] We rule with moderate strictness, and in return we are satisfied with moderate obedience. And I think that I can claim that our people are moderately sober, moderately chaste and moderately honest.'

1. AGEING

2   When he was writing his notes, which he left with the title *Senilia*, Schopenhauer was experiencing a period of belated satisfaction, because after years of Hegelian ostracism, his philosophy was gaining recognition. This permitted him to face his old age, the last eight years of his life, with a fighting spirit, attacking his adversaries. He wrote with clarity on many subjects and defended his theories rigorously.

3   David A. Snowdon et al., 'Linguistic ability in early
life and longevity: Findings from the Nun Study', in J.
M. Robine, B. Forette, C. Franceschi, M. Allard (eds.),
The Paradoxes of Longevity, pp. 103–113. As mentioned
in the main body of the chapter, the biographies of the
young novices, who were followed until their deaths (the
ages of which ranged from 78 to 97) revealed how being
full of enthusiasm for life in youth tends to translate into
a long and happy old age. Seventy-four nuns left their
brains to science, and the anatomic pathology studies
carried out showed how those who demonstrated a poverty
of language when they were young were more likely to
develop dementia in their old age. If you're interested in
longevity studies, I recommend reading the whole book,
where you'll find the chapter I refer to.

4   https://www.ons.gov.uk/
peoplepopulationandcommunity/populationandmigration/
populationestimates/articles/overviewoftheukpopulation/
july2017

5   Thomas B. Kirkwood, 'Evolution of ageing', *Nature*,
vol. 270 (24 November 1977), 301–304.

6   Leonard Hayflick and Paul S. Moorhead, 'The serial
cultivation of human diploid cell strains', *Experimental Cell
Research*, vol. 25 (1961), 585–621.

7   Ergün Sahin and Ronald A. DePinho, 'Linking
functional decline of telomeres, mitochondria and stem
cells during ageing', *Nature*, vol. 464 (25 March 2010),
520–528.

8   Helen Lavretsky et al., 'A pilot study of yogic meditation for family dementia caregivers with depressive symptoms: Effects on mental health, cognition, and telomerase activity', note 151, *International Journal of Geriatric Psychiatry*, vol. 28, no. 1 (January 2013), 57–65.

9   Giovanni Vitale et al., 'Low circulating IGF-1 bioactivity is associated with human longevity: Findings in centenarians' offspring', *Aging*, vol. 4, no. 9 (September 2012), 580–589.

10   Denham Harman, 'Aging: A theory based on free radical and radiation chemistry', *Journal of Gerontology*, vol. 11 (July 1956), 298–300.

11   Giovanna De Benedictis et al., 'Mitochondrial DNA inherited variants are associated with successful aging and longevity in humans', *The FASEB Journal*, vol. 13, no. 12 (September 1999), 1532–1536.

12   Julie A. Mattison et al., 'Caloric restriction improves health and survival of rhesus monkeys', *Nature Communications*, vol. 8 (17 January 2017), 14063.

13   Claudio Franceschi et al., 'Inflamm-aging: An evolutionary perspective on immunosenescence', *Annals of the New York Academy of Sciences*, vol. 908 (June 2000), 244–254.

14   Claudio Franceschi, 'The network and the remodeling theories of aging: Historical background and new perspectives', *Experimental Gerontology*, vol. 35, nos. 6–7 (September 2000), 879–896.

2. WHEN DO WE BEGIN TO AGE?

15   Claudio D'Addario et al., 'Transcriptional and epigenetic phenomena in peripheral blood cells of monozygotic twins discordant for Alzheimer's disease, a case report', *Journal of the Neurological Sciences*, vol. 372 (15 January 2017), 211–216.

16   Gian Paolo Ravelli et al., 'Obesity in young men after famine exposure in utero and early infancy', *The New England Journal of Medicine*, vol. 295, no. 7 (12 August 1976), 349–353. If you would like to know about the terrible 'winter of hunger', I can recommend the book by C. Carl Pegels, *The Second World War in the Netherlands: Memoirs* (2016), available as an ebook. Pegels was born in Holland and emigrated to the United States, where he became Professor of Management at the University of Buffalo. As a 7-year-old child, he witnessed the Nazi invasion of his village, near Rotterdam, and the fifth terrible year of occupation and famine.

17   Susanne R. de Rooij et al., 'Prenatal undernutrition and cognitive function in late adulthood', *Proceedings of the National Academy of Sciences of the United States of America*, vol. 107, no. 39 (28 September 2010), 16881–16886.

18   Daniel W. Belsky et al., 'Quantification of biological aging in young adults', *Proceedings of the National Academy of Sciences of the United States of America*, vol. 112, no. 30 (28 July 2015), E4104–E4110.

19   www.framinghamheartstudy.org

20 *Psychotherapy*, 2017.

21 Markus H. Schafer and Tetyana P. Shippee, 'Age identity, gender, and perceptions of decline: Does feeling older lead to pessimistic dispositions about cognitive aging?', *The Journals of Gerontology: Series B*, vol. 65B, no. 1 (1 January 2010), 91–96.

22 Rachel Pruchno, 'Aging: It's Personal', *The Gerontologist*, vol. 57, no. 1 (1 February 2017), 1–5.

23 In the chapter on work, Bertrand Russell explains how the modern man often fills his time working more than necessary, because he wouldn't know what else to do. *The Conquest of Happiness* was written when Bertrand Russell was 58 years old, and in the book, the philosopher admits that he is happier and enjoys life a lot more than when he was younger.

24 Hugo Westerlund et al., 'Effect of retirement on major chronic conditions and fatigue: French GAZEL occupational cohort study', *BMJ*, vol. 341 (2010), c6149.

25 Marie-Noël Vercambre et al., 'Self-reported change in quality of life with retirement and later cognitive decline: Prospective data from the Nurses' Health Study', *Journal of Alzheimer's Disease*, vol. 52, no. 3 (2016), 887–898.

26 http://thelistenersclub.com/2017/05/15/der-rosenkavalier-renee-fleming-and-the-passing-of-time/

27 The book caused quite a scandal, and, after Wilde's

death, James Joyce wrote a very respectful article in Italian which was published in the Trieste newspaper *Piccolo della Sera* on 24 March 1909, which ended: 'A verse from the book of Job is cut on his tombstone in the poverty-stricken cemetery at Bagneux. It praises his facility, *eloquium suum* – the great legendary mantle which is now divided booty. Perhaps the future will also carve there another verse, less proud but more pious: *Partiti sunti sibi vestimenta mea et super vestem meam miserunt sortis.*'

28   Giacomo Leopardi, *Zibaldone: Pensieri di varia filosofia e di bella letteratura* [The Notebooks of Zibaldone], vol. IV.

3. APPLE, PENNY, TABLE...WHY IS OLD AGE SO FRUSTRATING?

29   Alois Alzheimer, 'Über eine eigenartige Erkrankung der Hirnrinde' ['On an unusual illness of the cerebral cortex'], *Allgemeine Zeitschrift für Psychiatrie und psychisch-gerichtliche Medizin*, vol. 64 (1907), 146–148.

30   Ulrich Müller et al., 'A presenilin 1 mutation in the first case of Alzheimer"s disease', *The Lancet Neurology*, vol. 1, no. 2 (February 2013), 129–130.

31   Utermohlen, William, *Portraits from the Mind: The Works of William Utermohlen, 1955–2000.*

32   Juvenal, *Satires*, trans G. G. Ramsey (1918), http://www.tertullian.org/fathers/juvenal_satires_10.htm

33   Speech at the Enrico Greppi Award ceremony, Italian
Geronotology and Geriatrics Society, Naples, 1999.

34   Ronald C. Petersen et al., 'Mild Cognitive
Impairment: Clinical Characterization and Outcome',
Archives of Neurology, vol. 56, no. 3 (March 1999), 303–308.

35   Alan Morris, 'Gut microbiota: Link between the gut
and adipose tissues', *Nature Reviews Endocrinology*, vol. 13
(July 2017), 501.

36   Maddalena Muzio Treccani and Mario Rivardo, *Il
disegno e i colori nella clinica dello psicoanalista*, pp. 195–227
and pp. 221–252. As the authors explain in the preface
(Vol. I), 'the year spent running an art lab in a nursing
home in Milan […] focused on delineating psyche,
life form and its many variations in the relationship
each individual established with coloured pens and
pencils on a sheet of paper.' See also: http://www.
psicoanalisidifrontiera.it

37   Moyra Jones, *Gentle Care: Changing the Experience of
Alzheimer's Disease in a Positive Way* (Burnaby, BC: Moyra
Jones Resources, 1996).

38   Lorena Perrone and William B. Grant, 'Observational
and ecological studies of dietary Advanced Glycation
End products in national diets and Alzheimer's disease
incidence and prevalence', *Journal of Alzheimer's Disease*,
vol. 45, no. 3 (2015), 965–979.

39 Joshua Z. Willey et al., 'Leisure-time physical activity associates with cognitive decline: The Northern Manhattan Study', *Neurology*, vol. 86, no. 20 (17 March 2016), 1897–1903.

40 Brenda Hanna-Pladdy, Byron Gajewski, 'Recent and past musical activity predicts cognitive aging variability: Direct comparison with general lifestyle activities', *Frontiers in Human Neuroscience*, vol. 6 (July 2012), 198.

41 You can hear David Lang's symphony for 'Simple Song Number 3', sung by soprano Sumi Jo, on *Paolo Sorrentino – Music for Films* (CD, Warner Music Italy, 2017).

4. ACTIVE AGEING

42 World Health Organization, 'Active Ageing: A policy framework'.

43 Marco Pahor et al., 'Effect of structured physical activity on prevention of major mobility disability in older adults: The life study randomized clinical trial', *JAMA*, vol. 311, no. 23 (18 June 2014), 2387–2396.

44 Irwin H. Rosenberg, 'Sarcopenia: Origins and clinical relevance', *Journal of Nutrition*, vol. 127, no. 5 (1 May 1997), 990S–991S.

45 Matteo Cesari et al., 'Sarcopenia and physical frailty: Two sides of the same coin', *Frontiers in Ageing Neuroscience*, vol. 6 (July 2014), 192.

46 Jürgen Bauer et al., 'Evidence-based recommendations for optimal dietary protein intake in older people: A position paper from the PROT-AGE study group', *Journal of the American Medical Directors Association*, vol. 14, no. 8 (August 2013), 542–559.

47 In Sergio Rubin and Francesca Ambrogetti, *Pope Francis: Conversations with Jorge Bergoglio*, trans. Laura Dail Literary Agency (London: Hodder & Stoughton, 2013). The name 'Babette' was inspired by American writer Alice Babette Toklas, life partner of Gertrude Stein – who, like Babette, rose to international fame thanks to a book of recipes she published in French in 1954. Karen Blixen published 'Babette's Feast' in 1950, writing in English under the pen name Isak Dinesen, before she translated it herself into Danish so that it could be included with another four stories in the *Skæbne-Anekdoter* collection (published in English as *Anecdotes of Destiny*).

48 Sonja Hilbrand et al., 'Caregiving within and beyond the family is associated with lower mortality for the caregiver: A prospective study', *Evolution and Human Behaviour*, vol. 38, no. 3 (May 2017), 397–403.

49 https://www.mpib-berlin.mpg.de/en/research/lifespan-psychology/projects/the-berlin-aging-studies-base/base

50 Luisa de Vivo et al., 'Ultrastructural evidence for synaptic scaling across the wake/sleep cycle', *Science*, vol. 255, no. 6325 (February 2017), 507–510.

51   Gayle Doll, 'Bedroom scenes: Filmic portrayals of intimacy and sexuality in long-lived relationships', *The Gerontologist*, vol. 57, no. 1 (1 February 2017), 145–146.

52   Ruth Ray Karpen, 'Reflections on women's retirement', *The Gerontologist*, vol. 57, no. 1 (1 February 2017), 103–109.

53   http://cordis.europa.eu/project/rcn/206414_it.html

54   Figures collated by the United Nations: http://www.un.org/en/development/desa/publications/world-population-prospects-2015-revision.html

55   Claudio Franceschi et al., 'Centenarians as a 21st century healthy ageing model: A legacy of humanity and the need for a world-wide consortium (WWC100+)', *Mechanisms of Ageing and Development*, vol. 165, part B (July 2017), 55–58.

56   Craig R. Allen et al., 'Managing for resilience', *Wildlife Biology*, vol. 17, no. 4 (2011), 337–349.

57   Daniela S. Jopp, 'Physical, cognitive, social and mental health in near-centenarians and centenarians living in New York City: Findings from the Fordham Centenarian Study', *BMC Geriatrics*, vol. 16, no. 1 (5 January 2016).

58   Arnheim's diagram appears in *New Essays on the Psychology of Art* (Berkeley: University of California Press, 1986), p. 285. He says, 'Our way of looking at the seasons of human life is determined by two conceptions, which I

have tried to symbolize in a diagram (Figure 45). One of these conceptions is biological. It describes an arch rising from the weakness of the child to the unfolded powers of the mature person and then descending toward the infirmity of old age. In this view, the late style of life is that of the old man leaning on his cane – the three-legged creature, as the riddle of the sphinx describes him. It is the season of the "winter of pale misfeature," as Keats has it in his sonnet. [...] Correspondingly, there is another way of looking at the accomplishments of the ageing mind. This second conception complements the first by finding in the passing of the years an ever-continuing increase in wisdom. In my diagram, the symmetry of the biological arch is overlaid by a flight of steps leading from the limitations of the child to the high worldview of those who have lived long and seen it all.'

59 'Alla ricerca dell'elisir di lunga vita' ['In search of the elixir of long life'], *La Stampa*, 11 March 2015.

60 Claudio Franceschi et al., 'Inflammaging and anti-inflammaging: A systemic perspective on aging and longevity emerged from studies in humans', *Mechanisms of Ageing and Development*, vol. 128, no. 1 (2007), 92–105.

61 Jocelyn Kaiser, '"Rejuvenation factor" in blood turns back the clock in old mice', *Science*, vol. 344, no. 6184 (9 May 2014), 570–571; Saul A. Villeda, 'Young blood reverses age-related impairments in cognitive function and synaptic plasticity in mice', *Nature Medicine*, vol. 20, no. 6 (June 2014), 659–663; Steven M. Paul and Kiran Reddy, 'Young blood rejuvenates old brains', ibid., 582–583.

62 Joris Deelen et al., 'Genome-wide association meta-analysis of human longevity identifies a novel locus conferring survival beyond 90 years of age', *Human Molecular Genetics*, vol. 23, no. 16 (15 August 2014), 4420–4432.

63 Yi Zeng et al., 'Novel loci and pathways significantly associated with longevity', *Scientific Reports*, vol. 6 (25 September 2016), 21243.

64 Steve Horvath, Beate R. Ritz, 'Decreased epigenetic age of pbmcs from Italian semisupercentenarians and their offspring', *Aging*, vol. 7, no. 12 (2015), 1160-1169.

65 In 1850, Ludwig Feuerbach wrote an essay in praise of Jacob Moleschott's treatise of nutrition for the people, theorizing that mind and body are indivisible. A former pupil and critic of Hegel, Feuerbach was an anthropologist and humanist himself, and underscored the psychophysical totality of the individual and the fact that – to improve the spiritual life of a people – material conditions, starting from nutrition, must first be improved.
English translation by Cyril Levitt available at: http://citeseerx.ist.psu.edu/viewdoc/download?doi=10.1.1.456.2161&rep=rep1&type=pdf

66 Davide Gentilini et al., 'Role of epigenetics in human ageing and longevity: Genome-wide DNA methylation profile in centenarians and centenarians' offspring', *AGE*, vol. 35, no. 5 (October 2013), 1961–1973.

CONCLUSION

67 John Ruskin dedicated a whole chapter to Mount Cervino (the Matterhorn) in the fourth volume ('Of Mountain Beauty') of an immense, five-volume work called *Modern Painters*, written between 1843 and 1860 – a defence of J. M. W. Turner, who he considered the most talented landscape painter ever.

68 Martin Heidegger, *The Question Concerning Technology, and Other Essays*, p. 27.

69 Giuseppe Verdi, *Autobiografia dalle lettere*.

70 Stephen Cave, *Immortality: The Quest to Live Forever and How It Drives Civilization*.

Index

INDEX